D0597121

LIVES OF A BIOLOGIST

LIVES OF A BIOLOGIST

ADVENTURES

IN A CENTURY OF

EXTRAORDINARY

SCIENCE

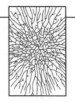

JOHN TYLER BONNER

HARVARD
UNIVERSITY PRESS
Cambridge, Massachusetts
London, England   2002

Pages 18–20 were originally published as the foreword to *On Growth and Form: Spatio-Temporal Pattern Formation in Biology*, ed. M. A. J. Chaplain, G. D. Singh, and J. C. McLachlan (New York: John Wiley & Sons, 1999); reproduced with permission.

Library of Congress Cataloging-in-Publication Data

Bonner, John Tyler.
  Lives of a biologist : adventures in a century of extraordinary science / John Tyler Bonner.
      p.   cm.
  ISBN 0-674-00763-8 (alk. paper)
  1. Bonner, John Tyler.   2. Biologists—United States—Biography.
  3. Biology.   I. Title.

QH31.B715 A3 2002
570'.92—dc21   2001051526
[B]

CONTENTS

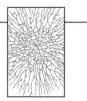

My lovely old office in Guyot Hall in Princeton—
which I have occupied for over fifty years—was
built in 1910. It has thick oak tables, brown wood
window frames, and four lights that hang from the high ceiling, each with a conical green glass shade. The only thing I
have added are bookcases on three walls. It definitely has a
beginning-of-the-century atmosphere.

At this moment I am sitting in my study in our summer
place in Cape Breton. This nineteenth-century house looks
out onto the estuary of a beautiful river. The walls of the
room are made of horizontal boards of uneven width that
have been smoothed with a draw knife. As I look out of the
window by my desk, I can see the occasional small boat on
the water and patches of farmland on the distant hills. Right
this minute all is enveloped in a morning mist.

These two rooms, in some sort of symbolic way, have
framed me for the century. Within that frame there have been
enormous changes. Here I am with my laptop computer
writing this, and periodically entering the Internet for references or for reading the latest in the *New York Times*. Biology, the world, and I have gone through many transmutations.

I was born in 1920 and started my interest in biology in the
early 1930s, which means that I have spanned a large portion
of the last century—a century that, above all things, has
been distinguished by an extraordinary series of advances in
biology. If one compares the biology of 1900 with that of
2000, the change has been astounding. In the beginning there
was the discovery of the genes on the chromosomes, followed

by the discovery of embryonic induction; the interpretation of evolution in terms of changes in gene frequency in a population; the rise of an understanding of the biochemistry of the cell; and perhaps most important of all, the beginning of molecular genetics, which was followed by the molecular genetics of embryological development; the great, new insights into animal behavior; sociobiology; the simplification of ecological and evolutionary principles by means of mathematical models—and so many other discoveries.

To one glancing at this history from a panoramic point of view, a number of things are evident. One is the pursuit of the gene and the extraordinary success of the totally reductionist approach, where the molecules and their structure have been unmasked. Another is the application of that reductionist knowledge to the problems of evolution and developmental biology. A third is the idealization, the simplification, of the vast number of complexities that characterize practically all of what we now know in modern biology: for instance, the regulation of the molecular activities within the cell, the complex structure of ecological communities, the functioning of the brain, the sequential steps of development, the patterns of animal behavior, and the intricacies of evolutionary change.

I ask myself where I fit in with all this. Over the years I have tried to keep my eye on as big a picture as possible, and to see how it all comes together. Development is not divorced from evolution nor, for that matter, from behavior; genetics is intimately intertwined with every aspect of modern biology from the activities of the cell to those of the brain; the grouping of organisms—either as social groups or ecosystems—are intimately connected with considerations of development and evolution.

One of the things that has happened during this hundred-year span is the enormous accumulation of new and fascinating facts—so much so that it is hard to see even the trees, let alone the forest; one could almost say that we are now focused on spots on the bark. A primeval urge inside me pushes me toward bucking this trend. However, I have not done it by simply dissolving into generalities, but have steered a quite specific path, a specific vision.

As will be evident in the pages to come, that vision has been to consider all organisms as life cycles, an enormously unifying way of thinking of things for it automatically links development, evolution, behavior, genetics, physiology, ecology, even behavior; it is a thread that links all of biology. The concept of life cycles also lies at the foundation of my own evolution.

I began with an interest in lower organisms, algae and fungi and related forms that confronted me with their trajectory from egg to adult, or spore to adult, in a way that could not be dismissed. Furthermore, my initial laboratory focus was on the development of the social amoebae (cellular slime molds) whose life cycle, as we shall see later, is so peculiar that it is impossible to avoid becoming conscious of life cycles. They have been for me the lens for all my work, both in the laboratory and in my writing: the life cycle is a great simplifying and unifying part of all living organisms. In particular it is the bond that connects development and evolution, subjects so central to all of biology and ones that have been receiving increasing current interest.

And then I am conscious of my own life cycle. Here I am

sitting in front of my computer—one complex machine facing another—and I am suddenly overcome by the chilling thought that at one moment in my life I was but a single cell, a fertilized egg. As my development unfolded through a combination of genetic and environmental instructions, I slowly, and sometimes painfully, became what I am today. This is my life cycle and it is another reason to see the world around me in those terms.

I am going to divide the century into five periods of twenty years and consider each from three points of view: one will briefly remind the reader what was happening in the western world; the second will show how biology advanced; and the third will discuss my own progress as a biologist and how I did and did not blend with all that was going on around me. It will be an eclectic mixture of history, biology, and autobiography.

Obviously the latter will be a bit difficult for the twenty years before I was born and to some degree for the 1920s, when I was a child. Nevertheless I have many reminiscences and tales of my parents and grandparents, who lived through that period, as well as some of the great men of biology, whom I got to meet or know later in their lives. From the 1930s on my involvement in the story will increase progressively.

As is true for everything I write, I am greatly indebted to a multitude of people for encouraging me and guiding me in my path. This book began six years ago when I wrote some memoirs for family and friends. A few told me I should rewrite them for publication and urged me to do so. I would like to thank Emily Wilkinson, Eric Rohmann, and Tory Prior for their advice on how to do it, much of which I have followed in these pages. However, I was too close to the original to think of working on it again, so I tucked it away and went on to other things. Much later it came back into my consciousness and I looked over the suggestions with a fresh mind and attacked it with a new spirit. Like all my writing it did not come into full bloom from the start but went through many drafts, and at each stage I received enormous help from many. Rebecca Roberts was the first to give me good ideas for improvement, and others followed, including Michael Fisher, two anonymous readers, and Slawa Lamont. I want to give my special thanks and appreciation to Jonathan Weiner. He was not only kind enough to read my efforts at every stage, but his mixture of wise suggestions skillfully coated with encouragement really made this book possible. It is my good fortune to have the help and support of such a superb writer and good friend.

Margaree Harbour
Cape Breton, Nova Scotia

LIVES OF A BIOLOGIST

It was a family joke that my father knew everyone, and if the name of some prominent or public figure was mentioned we could count on him to say, "Oh yes, I know Groucho," or "Alfred" (Lunt) and "Lynn" (Fontanne), or whatever the name. One evening a number of us younger ones were discussing Trotsky. My father looked up from his book and said, "Oh, I knew Trotsky." We all pounced on him saying, "This time, Pa, you have gone too far!" He said not at all, and that as a young man he worked for his father as a teller in Granddad's bank in Brooklyn, and Trotsky used his window. Trotsky had taken refuge in Brooklyn, and patronized his bank. Pa went on to say that they had quite lengthy discussions through the grill, and, knowing my father, I do not doubt the truth of this. I just wish he had written it all down after the meetings.

He had a remarkable ability to wrest easy conversation and information out of anyone. He would fix you with his large brown eyes, ask a question, and then manage to give the impression that every word one uttered was the most fascinating, the most precious of things. He could do that even with his own children. I can remember when I was a graduate student, his zeroing in and asking me about my research. He was not a scientist himself and I gave him a rather casual and superficial answer. He then said, "That isn't what you told me a year ago. Then you said . . ." He had remembered every detail of my experiments, and indeed things had changed in the intervening year, and he was not satisfied until he heard exactly what the new experiments had produced. He had such a remarkably retentive memory that I do not doubt if we had seen one another more frequently during that period he would have known more about my research than I did. That gift of using his large brown eyes to make you feel fascinating was no sham. He drank in everything and stored it away so that whether you were Trotsky or an offspring he could continue a conversation begun many months earlier without missing a beat.

I have often thought when advising gifted students—students who did everything easily and well—that they were at a great disadvantage. For them which courses to take, or what careers to pursue, presented momentous and impossible choices. All courses and all careers were desirable, and to have to choose was nothing but cruelty. My father was a person who chose them all and made a success of each. (One of my children in his early teens told me he did not want to discuss his future with me. When I asked him why, he said, "You wouldn't understand—you always knew what you wanted to do.")

Pa's first love was singing. He began as a choirboy in Brooklyn and continued in the glee club at Harvard. He went there because his father thought he should: Granddad was the sort of paterfamilias that no longer exists. I never knew him (he died when I was two); my impression has always been that he was rather like Clarence Day's father. He was autocratic, a bit eccentric, amusing, beloved by all the family, and definitely in charge. My grandmother Bonner, whom I got to know well, was very different. She was warm, motherly, and accommodating, with a well-hidden stubborn streak. I always thought of her as fitting the adjective "ample." She was forever embracing her grandchildren by surrounding them with masses of soft flesh, something we thought rather excessive, although we all loved her.

It is a dark secret, and I have never known exactly what happened during that freshman year in 1912. All we ever heard from Pa was that his father had yanked him out at the end of his first year at Harvard because he was going out with what he quaintly called "chorus girls." I have always had a feeling that this glossed over a good story. All his life he retained his interest in pretty women—those big brown eyes were busy until almost his dying days. Sometime after this initial foray he ended up imprisoned behind the bars of the teller's window, to escape only by talking to Trotsky and no doubt many others.

What he really wanted to do was sing. He had a fine tenor voice and asked his father if he would pay for lessons, preferably in some place far removed from Brooklyn, far away from the bank. The answer was an outraged "no." He was not to be deterred, and by some long-forgotten stratagem he managed to persuade Enrico Caruso to give him an audition. That great star of the Metropolitan Opera was so favorably impressed that he said he would speak to my grandfather. The latter was partially subdued by this flank attack with the result that, with the financial help of a great-aunt, my father went to Italy to study with a well-known basso recommended by Caruso. Those two years must have been glorious, and he often told us about them when we were young. In his youthful energy he did all sorts of pranks with his spirited friends, pranks that today seem rather antique. For instance, they carried a suitcase full of horse manure to Venice, carefully placing it on the Piazza San Marco in a neat pile in the middle of the night. They stayed up all night to watch at dawn the first startled Venetian who must have thought the famous horses on the cathedral had swooped down from above. The thing I remember most vividly of his Italian stories was his charismatic teacher telling him that "you have not lived unless you have loved and starved." As a small child this seemed to me a wonderfully romantic thought, but as I grew up I realized that while my father had undoubtedly managed the love, he never came close to starving—yet he certainly had "lived."

The climax of these two years of study was a contract to sing in Italy: he was going to achieve his greatest dream. But it never happened, because as he was waiting for his initial role the First World War broke out and he had to return home to New York City. There he met and courted my mother, and just after they were married he joined the army and went to boot camp with all its drudgery. Yet even here he managed to end sunny-side up: he rose from private to second lieutenant and ended up in an intelligence-counterspy unit in France. This period produced no end of riveting stories for young children, such as going in disguise into Paris opium dens and tracking down criminals involved in drug pushing. Today this has all too familiar a ring, but at that time it was a new threat to the well-being of our soldiers. The climax came when he single-handedly arrested a kingpin in the trade by reaching into his jacket pocket as he was shooting pool and holding him up with his own gun. He was very good at telling these stories and filled them with suspense so that his small sons were wide-eyed with excitement and pleasure.

When he returned from France in 1919 he was greeted by my mother and their first child, Paul Jr., whom he had never seen. They settled down in New York, but singing was out of the question—he now had a family to support.

The world was changing fast during those twenty years. I can remember my father saying he had a vivid recollection of the enormous excitement of seeing his first automobile shortly after the turn of the century. Marconi's wireless radio began its flowering as did many of the inventions of Edison. The period began with ubiquitous top hats and derbies and stiff collars for the men, and long dresses with choke collars, whalebone corsets, and enormous hats for the women, all of which were to disappear by the 1920s. The horrors of the First World War had much to do with the changes—everyone was shaken to the roots. And there was the Russian Revolution whose effects extended throughout the century and whose repercussions are still with us today. What did Trotsky and my father talk about through the grill?

My mother was born and spent the first fourteen years of her life in Zürich, Switzerland. Her father was the younger brother of a family of successful silk weavers and had come to America to start a United States branch. My grandfather Stehli was another man of his time, in this case a Swiss version. His concept of the family and the roles of the husband and wife and the children fell into well-defined slots. He was the boss in all matters, and since he was an intelligent and kind man, his orders, even the stern ones, were followed without question or argument. His older brother, who ran the business in Zürich, seemed to us a far more severe person, but they both ran their families in the same way, and no one raised a question. For instance, when it was announced that the Kaiser had invaded Belgium to begin the First World War, my mother told me he came down to breakfast in a cold fury and announced that henceforth German would never be spoken in the family again—and it was not. Swiss-German, English, and French became the official languages.

His was an era when the double standard was accepted (at least by the men), and as a moderately rich European Grandpa it was unremarkable that he should (discreetly) have a mistress, but philandering was out of the question for my grandmother. There were mutterings about Grandpa's illegitimate son, but no one thought much about it—it was a matter of course. All these family and nonfamily matters were part of the time; they do not say anything about him.

He was a warm and generous person. He could quote Goethe and Schiller and other German poets at length, and he was, before I was old enough to hear him, a superb violinist. In his younger days he had played as an amateur among the first violins in the Zürich orchestra. He told me he had heard Brahms conduct there, something I found hard to believe—surely he was not that old! When I was very young he developed a severe tremor in his hands and had to give up playing. When he would come to visit us in New Hampshire he would ask me to play him some of my records. One day I put on a Schumann quintet that he said he used to play, and he sat there with tears streaming down his cheeks.

The first twenty years of the century were of great importance to the flowering of biology that was to continue its steep upward sweep right to today. It began with the well-known rediscovery of Gregor Mendel's paper on genetics. The world had not been ready for it when it was first published in 1866, but the paper finally came to light in 1900. It was like finding a key piece of a puzzle that had lain hidden, out that, once in place, had an electric effect. Mendel showed that the characters of the pea plants that he grew in his monastery garden sorted out in such a way in his breeding experiments that they seemed to be controlled by independent factors, which, after the turn of the century, came to be known as genes. Very soon it was realized that these genes must be carried on the chromosomes in the nuclei of cells, and all the beautiful work done on cell structure in the nineteenth century, especially by the German cytologists, suddenly made sense in terms of the new science of genetics. The elaborate activity of the chromosomes during cell division, and especially in the divisions making the sex cells and their fusion in fertilization, could now be interpreted in terms of the carrying and the intermixing of Mendel's unit characters. As we will see, as one proceeds through the twentieth century the influence of these discoveries has been enormous—it became what Evelyn Fox Keller aptly calls "the century of the gene." Each time Mendel's genetics collided with other biological entities, as it has, it produced a flash of lightning, and stunning progress followed. The first is the one just described, which was to be cytogenetics, the next in the 1930s was the fusion of genetics with evolution in the form of population genetics, followed by the fusion of genetics with biochemistry to start molecular biology in the 1950s, which led to the current great interest and enormous progress in the genetics of animal and plant development.

One of the key players in the early establishment of genetics was Thomas Hunt Morgan. He and his remarkable students and associates, in some very clever experiments, showed that the genes were in a linear order on the chromosomes; furthermore, they could actually map where the genes were on the chromosome. These earth-shaking experiments, which all began with work on fruit flies in a small room at Columbia University, had an enormous impact, and Morgan was appropriately awarded the Nobel Prize.

Many years later, after my freshman year at college, I spent a summer at the Marine Biological Laboratory in Woods Hole and met Morgan and a number of the other giants of his generation. After taking the botany course I asked to stay on to do some research, which my faculty mentors agreed to. All I needed now was parental permission, which was readily given, but my father added that he was coming down from New Hampshire to visit for a day or so. Both bits of news delighted me, and there is one special reason I shall never forget his stay.

We took a walk out to Penzance Point from the Laboratory, and at the very end came to Dr. Warbass's property at the tip. He was an ancient, retired obstetrician from Brooklyn, and he had a big sign at the end of the road welcoming visitors to walk around the point on his land, but to please stay on the path. We no sooner accepted this invitation when we encountered Dr. Warbass cutting hay in a field. He stopped and leaned on his scythe, making quite a figure with his long, flowing gray beard. I had never met him, but felt compelled to say something, so I thanked him for letting us walk on his path, and then said, "I would like to introduce my father, Mr. Paul Bonner." Dr. Warbass nodded and said, "Are you from Brooklyn?" and my father replied that indeed he was. Then Dr. Warbass said, "Was your mother's name Theodora?" Again my father, somewhat surprised, said yes, at which point Dr. Warbass managed a memorable coup de grace by saying, "I delivered her." The only thing that was missing was an hourglass beside his scythe.

Professor George Parker, one of the grand old men of physiology, was lecturing that evening at the Laboratory, and fortunately I persuaded Pa to come. Professor Parker must have been in his eighties and he was famous for having been one of the first to appreciate the fact that substances were secreted at the ends of stimulated nerves, and that these "neurohumors" could simulate the surrounding cells. He discovered this studying color changes in the skin of fish: the neurohumors caused the pigment cells to expand or contract. In his lecture he pointed out that this was also the mechanism of blushing in humans, where the neurohumor caused vasodilation of the small blood vessels. Dr. Parker was in rare form, and he finished by saying that in his youth he had always wanted to know how far down from the face a blush extended, but when he was young, women wore their dresses tight around the neck so he never could find out. He then added, with wonderful Victorian relish, "But I still don't know the answer, because now one can see almost everything, but women don't blush anymore!"

When I was an undergraduate at Harvard, Dr. Parker, then very emeritus, had his office on the floor above me. Upon occasion I used to go up there to chat with him, and although in some ways he was a distant sort of person he seemed pleased to talk to an eager young man and was at no loss for words. We chatted about current trends in physiology, especially his neurohumors and I know I benefited greatly. He was never overbearing, despite his impressive appearance, and had a fine, thinly disguised sense of humor. I was at the Marine Biological Laboratory one summer and Oscar

Schotté, a delightful and amusing star pupil of the famous Hans Spemann in Germany, came down from Amherst to give a lecture on parthenogenesis—embryonic development without fertilization—in the embryology course. He sat next to Dr. Parker at breakfast in the mess hall and said to him that he was in an absolute panic because he had accidentally left his notes behind. Parker said to him that what he should do is just "open his mouth and let the Lord speak." Schotté trundled off to the class and began by telling the story of his misplaced notes and Dr. Parker's advice. Then he said, "Now, I can hardly wait to hear what the Lord has to say about parthenogenesis."

For a young undergraduate, one of the most exciting aspects of Woods Hole was the panoply of Olympians, who to me seemed like jewels in a crown. Besides Professor Parker, there was R. G. Harrison who discovered how to grow animal cells outside of the body, E. B. Wilson the author of an enormously influential book on the cell, E. G. Conklin (of whom I will have more to say presently) who, along with Wilson, in early Woods Hole days discovered mosaic embryonic development (where in some invertebrate animals each part of the fertilized egg is destined to become a particular part of the adult), T. H. Morgan, the father of modern genetics, and numerous others equally distinguished. They all seemed to me ancient, which only reinforced their godlike image in my mind. Here were real "grand old men" of stature and accomplishment; I admired them all without reserve and felt enormously privileged to be in their midst.

A close friend of my parents was the nephew of T. H. Morgan, and when I told him I was going to Woods Hole he asked me if I would like a letter of introduction to his uncle. Of course I said yes, and in the middle of the summer slipped it into Professor Morgan's mailbox with a note asking if I could call on him. This was followed by utter silence, so I consulted my mother as a faultless arbiter of social mores. She said that it was quite proper to just pay a call, which I did with my heart near the roof of my mouth. It turned out to be a great success, for he and his charming wife and daughter could not have been kinder to this frightened youth. They made me feel at home, which must have taken some skill. Dr. Morgan was famous for being rather parsimonious in small things (although the opposite in big matters: he shared the money from his Nobel Prize with the key members of his laboratory). He told me that when his daughter was a small child she went up to Dr. Parker, who greatly resembled him, beard and all, and said, "Daddy, could I have a nickel for some ice cream?" Dr. Parker gave her a dime, which she carefully examined, and then said, "You're not my Daddy." Talking to an Olympian was a pretty heady experience for an embryonic biologist.

As will soon be evident, embryology, which later came to be called developmental biology, has always had a special fascination for me—in fact, I became a developmental biologist. The subject was already flourishing in the first twenty years of the century as it had in the latter part of the nineteenth. Again Germany had been the particularly important center for this progress, led by such figures as Wilhelm Roux and Hans Driesch and others, as well as the pioneers of plant development.

In many ways Roux was the founder of experimental embryology. His pioneer experiments in cutting and altering cells during the early development of amphibian embryos (that were easy to manipulate) were not only interesting in themselves but had an enormous impact on twentieth-century embryology. His basic innovation centered around the "mechanics of development" *(Entwicklungsmechanik)*, the modern idea that one stage or step of development was the mechanical *cause* of the next. Hans Driesch did some famous experiments with the equally cooperative embryos of sea urchins and on the regeneration of hydroids (relatives of jellyfish). He apparently was a man full of charm and liked by many, including T. H. Morgan who was a good friend. His only misstep was that after his brilliant career as an embryologist, he embarked on the slippery slope of philosophy. Among other things, he was so amazed by the remarkable ability of embryonic cells to recover after mutilation, as he discovered in his famous experiments, that he became a vitalist; that is, he believed that only some supernatural force could account for the incredible ingenuity of the development of an embryo that was able to repair itself after mutilation. It could not possibly be explained by Roux's developmental mechanics, and only some inner spirit, some mystical force, could account for the miracle of embryonic development. Vitalism was no more popular at the beginning of the twentieth century than it is today, but our admiration for his experimental work has not flagged.

One of the most important discoveries in the early part of the twentieth century was that of embryonic induction. Hans Spemann and Hilda Mangold showed that one part of an early developing newt stimulated another part it touched to develop into an embryo, and this they demonstrated by grafting an extra "organizer" region, as they called it, from another embryo, which resulted in a second embryo forming on the side of the embryo that received the graft. In America embryology also flourished early in the century. Particularly notable was the work mentioned earlier of E. B. Wilson, E. G. Conklin, R. G. Harrison, and their students.

One summer at Woods Hole some years later stands out in my memory for many reasons. My perceptions of the Marine Biological Laboratory had much changed since my teens, but I still held it in awe. I was a beginning assistant professor at Princeton where Dr. Conklin was an emeritus professor, and when he heard I needed a place in the Laboratory he asked me to share his room, which was given to him in perpetuity as one of the founding fathers of the MBL. He was in his eighties, and an extraordinary man. He had this forceful, strong voice, and the sentences came out of him as though he were preaching from the pulpit. It was a wonderful experience for me because I had always admired him, both for what he had done for the frontiers of embryology, particularly in discovering mosaic development, and for what he had done for biology at Princeton, where Woodrow Wilson had recruited him in the early part of the century to build a biology department. Now I was to get to know him.

My only problem was that he turned out to be intensely garrulous. The difficulty was that what he had to say was continuously fascinating, making it very hard for me to get any writing done. The only way I managed it was to disappear in the library and write there.

I still remember many of his tales. In the first place he knew a number of the great nineteenth-century German embryologists and he would tell me about them. I particularly remember that he disliked one of my heroes, Hans

Driesch, who had discovered regulative development, which seemed to be at odds with the mosaic development discovered by Conklin and E. B. Wilson. Simply put, in mosaic development, which is found in some invertebrates, if one cuts an embryo in two at an early stage, each half produces a half-embryo. Driesch discovered that in the embryos of other invertebrates, each half would regulate after the operation, and two small but perfect embryos be produced. I could never determine whether Conklin's antagonism toward Driesch was because of the apparent conflict between mosaic and regulative development or because of a personality difference. I suspect it was the latter, and furthermore, as I mentioned above, Driesch became a vitalist, which clearly Conklin considered an unhealthy occupation.

The things Conklin told me were not all embryology. When he heard I had spent World War II in and about Dayton, Ohio, he told me that as a young man he had been in a bicycle race that went through Dayton. It was called "The Century Run," and it was a course of one hundred miles. The bicycles were front wheelers, with a huge front wheel and a very small back one. He said none of the streets of Dayton were paved. I assured him Dayton had quite changed.

One day I was sitting alone in his room, and Dr. Otto Loewe, a charming and amusing Viennese, dropped in looking for Dr. Conklin. Loewe was famous for seeing a wonderful experiment in a dream that proved nerves give off chemicals in the blood that can affect distant tissues; it was an experiment that got him the Nobel Prize. Many have dreamed of getting the prize, but I believe he is unique in having a dream that gave him the prize. This, of course, is not at all fair to his great abilities, but it is partly his fault for it was his own self-deprecating humor that gave rise to the story. We started talking and he said to me, "Do you fully appreciate your enormous privilege of sharing the room with America's greatest embryologist?" I made the appropriate noises and he went on to say he had always been a great admirer of Dr. Conklin, and he always made a point of going to his annual lecture in the embryology course. "Have you noticed the way each year he ends the lecture so dramatically, saying in a quavering voice [which Dr. Loewe imitated with much pathos in his thick German accent], 'This may well be my swan song.' It's amazing the effect it has on the audience. You know, I'm not so young myself, so now every time I end a lecture I do the same thing with great effect. Soon I plan to publish a book called 'My Collected Swan Songs.'"

Biology in those first twenty years of the century was a mixture of many new ideas and discoveries that laid the foundation for all the progress of the rest of the century, combined with the tremendous overshadowing influence of the latter half of the nineteenth century. That had been an enormously important period. It produced Darwin who laid the cornerstone of evolutionary biology, Pasteur who did the same for biochemistry and microbiology by finding the chemical nature of fermentation and the bacterial nature of diseases among many other brilliant advances, and Claude Bernard who laid down the modern foundations for mammalian physiology. More than any other country, Germany was the breeding ground of great advances in biology: those in embryology I have already mentioned, and without making a great list of names, it was largely the German schools of biology that were responsible for establishing the morphology—and its changing—of cells during development, including the structure and behavior of the chromosomes. During that period the Germans also put botany on an elevated footing by establishing higher plant anatomy as a significant discipline and providing the groundwork for understanding plant development. I have not even touched the surface of the magnitude or the diversity of the advances in the nineteenth century. I remember reading an amusing reference to a Ph.D. thesis in history that the eager student had initially called "The Influence of the Fifteenth Century on the Sixteenth." Now that no longer seems a joke to me when I think of the enormous influence the biology of the nineteenth century had on the twentieth.

The period between 1900 and 1920 really was an exciting and brilliant extension of what had gone before. No doubt the most significant step right in the beginning was the fusion of Mendel's heredity experiments with what was known of the chromosomes in the egg and sperm and their fusion at fertilization. And there were many others, such as the evolution of experimental embryology already described, and all the great progress in our understanding of disease, the steady advances in biochemistry and physiology, plant development, and other disciplines. All of these achievements, big and small, arose from mining the riches laid at the feet of the twentieth century by the nineteenth century, and right from the beginning they were vigorously exploited.

Even the way science was done did not change much; those changes came later. All the equipment in the laboratory was of the utmost simplicity: crude balances, hand-cranked centrifuges, lovely brass microscopes (with fantastic optics developed by the early German optical companies), and the simplest kind of glassware. What innovations there were involved perfecting what already existed. To give an example, Hans Spemann was able to perform his delicate experiments on amphibian embryos by inventing simple ways to make glass microneedles for his operations. He also perfected a way of tying a hair loop around an early embryo to partially constrict it, which required fine jewelers forceps and the soft hair from a newborn baby. This is a perfect example of the technical advances of that period; it was not an era of high-speed centrifuges, electron and confocal microscopes, automatic balances, and all the other complicated, often computer-controlled, equipment that we use and need today. The apparatus then were extremely simple by comparison, but the ingenuity of their operators was no less clever and innovative than it is today.

Another figure of that period who had a great influence on me throughout my biological life was D'Arcy Wentworth Thompson, who published his famous book, *On Growth and Form*, in 1917. He was a professor of natural history at Dundee and St. Andrews in Scotland for an amazing span of sixty-seven years, and among his gifts was an impressive command of many languages, especially classical Greek. He translated Aristotle's *Historia Animalium* for the Oxford series of Aristotle's works and wrote a *Glossary of Greek Birds* and a *Glossary of Greek Fishes*, in which he gathered all the birds and fish mentioned in classical Greek literature and from his zoological knowledge illuminated the references, fusing science with the classics: who could do that today?

Much has been written about D'Arcy Wentworth Thompson and his famous book, the most recent being a symposium in his honor in 1999. *On Growth and Form* continues to be read—it has enjoyed really quite a remarkable record of publishing endurance, and while it cannot compete with Shakespeare or the Bible, in all fairness they had a head start.

I have often asked myself how can one explain this extraordinary longevity of Thompson's great work? The answer is not to be found solely in the increased interest today among biologists in the use of mathematics for understanding living phenomena, although that is certainly an important factor. But there are other reasons as well. In 1961 Cambridge University Press sent me a thick package of reviews of Thompson's book, and while the majority were from newspapers and the usual biological publications, there were reviews in specialized journals for psychiatrists, veterinarians, and orthodontists! But most interesting to me has been the fascination for the book by the art world: painters, sculptors, crafts workers of all kinds, and architects. It was even listed and highly recommended in the *Whole Earth Catalog*, that symbol of the revolutionary days of the 1960s and 1970s. There is something about form in nature that has a universal appeal.

I suspect that one of the most important reasons for its continued popularity is the beautiful prose. It is a sheer pleasure to read. Thompson's style has few equals; there is poetry in his sentences. Even when he says things that are no longer true, the reading of them still stands. It is hard to imagine, however, that this is the reason biologists still hold him in such high esteem. That I think comes more from his vision of how mathematics and physics can contribute to the form of living creatures. Nowadays, with the great flowering of molecular developmental biology, there is an increasing awareness that even though genes and the proteins they produce are enormously important, they alone cannot account for the growth and development of animals and plants. One must also consider the physical forces—which can be expressed in simple mathematical terms—for no matter how clever the activities of genes, they do not work in a vacuum. The substances that they create, and the environments in which they are created, are subject to physical laws and constraints that tell us much about the forms that are produced. Only part of the show is in the genes; the rest has to do with the properties of matter. Even though D'Arcy Thompson dismissed genetics, he, more than anyone, began our thinking about the other part of the equation, the appreciation of which is almost more relevant today than it was in 1917. So his book and his ideas survive with vigor because molecular developmental genetics has exposed what else is needed.

I became intimately familiar with the book when I rashly agreed to do an abridged edition in 1960. It was an enormously difficult job, largely because I knew I was tampering with an acknowledged classic. I was determined not to alter the prose, but somehow I had to retain the book's continuity while removing large sections. Once I started I soon saw my difficulties: it was like being asked to reduce *Hamlet* to a third of its length without rewriting a word. My problem was helped a bit by adding commentary here and there, but nevertheless I knew I was playing with fire. As one might expect, the reaction was mixed: many—especially those to whom the book was new—actually read the abridgement where they might otherwise have been overawed by the huge 1942 edition. At the same time many scholars, especially mathematicians, were furious at my presumption. A distinguished colleague in the mathematics department at Princeton, who I thought might be sympathetic, told me angrily that I had committed an unpardonable sin by removing most of the equations. Since the abridgement's publication in 1961 I have on two occasions given a lecture at St. Andrews University on my experimental work with cellular slime molds, and both times I was introduced in a funereal tone as "the man who abridged D'Arcy Thompson's *On Growth and Form*." Their great professor could manage quite well on his own—my help was not needed.

Clearly they have it right. The mere fact that there has been a recent symposium in his honor is evidence that Thompson's influence continues today. Biologists and mathematicians interested in living forms remain inspired by his message. Here is the essence of that message in his own words:

> Cell and tissue, shell and bone, leaf and flower, are so many portions of matter, and it is in obedience to the laws of physics that their particles have been moved, moulded and conformed . . . Their problems of form are in the first instance mathematical problems, their problems of growth are essentially physical problems, and the morphologist is, *ipso facto*, a student of physical science.

Unfortunately I never met D'Arcy Thompson. The first course I taught at Princeton in 1947 was an upperclass course in which I used his book. I wrote to tell him and got a very sad letter in reply. He had gone on a lecture tour in India that broke his health; he died shortly after at the age of eighty-eight. An older colleague of mine had been on the same lecture tour and said that D'Arcy Thompson's lectures were a wonder. He was an imposing, large man with a flowing beard, and apparently he gave one entire lecture on the vertebrate skeleton, all the while holding a very irritated chicken under his arm, pointing out its parts as he went along.

One particularly interesting aspect of this era was the skepticism about natural selection, or Darwinism, being an adequate explanation for evolution, and D'Arcy Thompson was among those skeptics. There was a quite general feeling that evolution was far too grand a process to be explained by such a simple and seemingly arbitrary process as natural selection; there must be another force or principle that pushes evolution forward in what seemed obviously to be some form of progress. This progressive force was sometimes called "orthogenesis," and while vitalism was rejected by most biologists then, they were often quite comfortable with the idea that there was a driving force that made organisms evolve into bigger and more complex forms. Such a view was compounded by the skepticism or lack of appreciation of Mendel's laws of heredity, and the more general acceptance of Lamarckism, the inheritance of acquired characteristics. This illustrates an interesting phenomenon in the sociology of science: we are greatly influenced by the fads and the fancies that surround us at any one time and it is hard to break out of the mold. Today we are in a new mold, but as I shall discuss much later on, we are still changing. However, even though for some years after 1920 there were some notable proponents of progress as being inherent in evolution, today they are pushed aside as an insignificant fringe.

In subtle ways my mother had an even greater influence on me than my father. She was a person of enormous internal strength, something that was not well hidden—it was obvious the moment she entered a room. She was a handsome rather than a beautiful woman. She always complained that her nose was too big, as were her hands, but I think those two features were the key to her attractiveness. She had a face full of character and her hands were strong and competent. Part of the reason she was so conscious of what she considered her deficiencies was that her mother, who had been a great beauty in her youth (and really remained so in her stately old age), was forever telling her daughter that she was somewhat imperfectly made, if not an ugly duckling. I think most women would have been permanently crushed by such continuous, unencouraging comments, but my mother had two great defenses against this and all other adversities. She discussed everything with her older children, and no doubt with others before we came along, so that she always had the therapy of telling someone rather than keeping it pent up. Much more important, however, was the simple fact that she was a person of extraordinary self-assurance; it just oozed out of her. That assurance gave her the freedom to enjoy amusing incidents, and not to feel inhibited in showing that pleasure.

As I look back at our relationship, I suppose all of these things were important to me, and no doubt I tried to emulate what I admired. But the real bond between us was my admiration of her quite exceptional intelligence. This manifested itself in her wide reading, and the way she understood what she had read. More to the point, she was intelligent about life, about people, about what was going on in the minds of her children. She knew exactly what to tell us (which was far more than most parents ever tell their offspring), but she never said the wrong thing, or stepped over the line she had drawn for herself. I can remember her telling me when I was in my teens, with amusement in her voice, that when she returned from her honeymoon, her mother, much to her surprise, immediately took her to one side to ask about all the details of what went on in her nuptial bed; she wanted to be sure all systems were "go."

I have only dim memories of my childhood. We lived in New York City and then in Long Island for my first ten years. I remember being very unhappy as a teenager, but I do not really remember why. My mother kept a diary from 1930 to 1935, but unfortunately she says nothing about herself and her life; it is all about the daily activities of her four boys (our ages spanned ten years), making limited reading, with entries such as "Henry vomited at lunch" and similar historic events. The annual summary pages are the most interesting: after the first year she says about me, "John is pig headed and argumentative, but strangely thoughtful and sweet. Very bright." Things are a bit better by 1935: "John has improved unbelievably. He has become articulate, much less shy, and his brain is working clearly and not in the former muddled fashion."

As must be clear from these diary fragments, Ma liked her children mainly because of their potential. She was impatient with childishness. When I would do something that was atavistic, she would fix me with a glacial stare and say, "I can hardly wait until you grow up." She did favor the older ones, and Paul and I received more attention. Tony, who was eight years younger than I, spent more years than any of us with a nanny.

I never liked nannies; on the whole they seemed quite horrifying to me. Just an extra someone to boss one about. I remember one in particular who was very strict and told me sternly that masturbating was bad for one—I would lose spinal fluid and become quite round-shouldered!

Ma had remarkable control over herself—she always said the right thing and always had the correct expression to go with it, a quality that enabled her to become a highly skilled poker player, a game she loved. My father was also an addict and a member of the Thanatopsis Club, which included their whole set of literati and Broadway friends, but they often played with a large group at home or at the houses of friends, and my mother could then join in and demonstrate her wonderful skills. When I look back at this now I am reminded of the Marquise de Châtelet, a brilliant mathematician and Voltaire's mistress, who, when short of cash, would descend on Versailles for a spot of cards, coming back with enough money to last her a long time.

Grandpa used to tell us to never play cards with strangers, but it was Ma who taught me the real lesson. She joined us for a family game of poker one evening—the stakes were appropriately low. She and I had a bit of a duel, and she absolutely flattened me. As she left she said that I must improve at the game before I played with others. I was furious in my humiliation, but it had the peculiar effect that I have never had any desire to gamble for the rest of my life. I so disliked that sinking, powerless feeling of losing that I never wanted to have it again. This helped me years later during the war when I was in the army. I was sent to Florida to test some equipment, and I shared my compartment on the train with a pleasant-looking young man my age, who asked me if I would like to play some poker. I said yes, but I would not play for money, which he thought amusing (and absurd) and asked me why. I explained that I was incapable of winning. After a period of silence he said to me that he had a confession to make: he was a professional gambler! He was on his way to a new job in a casino and he needed to keep his hand in; would I mind if he practiced? Naturally I said no, and he proceeded to deal about eight hands of blackjack, and then asked me to check him. He went around the circle and for each one he named the bottom, face-down card while I looked, and he had every one right. Quite unnerved I said, "I don't suppose you would tell me how you do this." He laughed pleasantly and said that I supposed correctly. I did ask him if the cards were marked, and he looked rather insulted, and said if I had a deck he would do the same with my cards. I have been playing solitaire ever since.

Life in Locust Valley, Long Island, in the late 1920s was, in my imagination of the distant past, exactly like the atmosphere created by F. Scott Fitzgerald in *The Great Gatsby.* The men wore white flannel trousers in the summer heat, and the women wore flowing dresses and wide-brimmed straw hats. Bootleggers in dark serge suits appeared in black cars carrying cheap suitcases full of spirits and French wines. Deadly croquet was played on the lawn in front of the house. The children had the woods, the ponds, and the marshes to plore. It was there that I first became fascinated with the out-of-doors, a fascination that later led me into biology.

My parents had an ever-increasing circle of friends, many of them in the worlds of journalism, literature, and the arts. Regulars were Franklin P. Adams (or F.P.A., who used to captivate us children by tapping out tunes with a pencil on his large white teeth), George S. Kaufmann and Mark Connelly the playwrights, Dorothy Parker, Edna Ferber, Harpo Marx, Alfred Lunt, Lynn Fontanne, and many others. Alexander Woollcott, one of the original contributors to the *New Yorker* and a central figure among the popular literati, was their closest friend, for in a kind, sterile sort of way he had a permanent crush on my mother. None of this made much of an impression on me, except when someone was nice to the children. I think the only time I was really impressed was when Harold Ross, the editor of the *New Yorker,* appeared and sat down in our living room for some time before he suddenly said, "Oh I forgot, I left someone in the car." That "someone" turned out to be Ginger Rogers. Even to me at that preglandular age she seemed so beautiful—how could one possibly forget and leave her in the car? It turned out they were childhood friends, but that explained nothing as far as I was concerned.

Ma was quite open about her fondness for Alexander Woollcott, just as he was about his feeling for her; it was a harmless affair. He once told her, and of course she told us, that he got more pleasure out of a good bowel movement than sex, which can hardly be described as a passionate declaration. He admired her intelligence, and she admired his wit, as indeed we all did. One of his standard jokes was how

he could successfully murder my father, but in fact th[e]
were good friends. When we lived in Locust Valley, Wool[l]
cott rented a cottage from my grandfather where he stayed
quite a few months. Once a week my brother and I would go
down and he would read to us for a couple of hours. He read
us good, sentimental children's stories about dogs and glory
in his mellifluous voice, and we loved it.

Woollcott told my mother that he thought I was intelligent, which she immediately repeated to me. I was flattered
then, but later I discovered it was a barrier between us. When
I pursued a university career I began to make him nervous. I
do not think I was either pedantic or arrogant (although perhaps I was), but he had both a fear and a faint dislike of someone who chose an intellectual life. He overcame it to some
extent, for the sake of my mother, but we never were on easy
terms after I had grown up. On my sixteenth birthday he
suddenly sent me a pair of gold cufflinks holding two fragments from a huge find of dinosaur eggs discovered by Roy
Chapman Andrews in the Gobi desert, and with it came a
characteristically amusing letter. It ends:

> I think you will be justified in regarding your cufflinks
> as antiques. Roy Chapman Andrews, who headed the
> aforesaid expedition and who gave the shell fragments
> to me in Peking some years ago, airily fixed their age as
> 90,000,000 years but I seem to detect in Mr. Wells (H.G.
> in the "Science of Life") a tendency to fix their age
> nearer 50,000,000 years. I am myself a conservative by
> nature so when I had to declare them to the custom's
> authorities at the port of San Francisco, I contented

BECOMING A BIOLOGIST

myself that they were "more than a hundred years old."

Dr. Andrews told me that the only other person to whom he had given similar pieces of the dinosaur shell was Jack Barrymore. Barrymore said that anyone who has been an actor as long as he had was accustomed to the thought of having old eggs thrown at him. All of which leaves me time and space only to wish you many happy returns of last Tuesday.   *A. Woollcott*

I was always sorry I made him uncomfortable.

Those dinosaur eggs press an automatic button in my memory. My brother Paul took a course at Harvard with Professor Langdon Warner on oriental art in which he described an archeological expedition to the Gobi desert. One day his caravan met with that of Roy Chapman Andrews, whom he knew to be there. Professor Warner went over on his camel to greet Andrews by saying, "Doctor Livingston, I presume," at which point Andrews drew himself up and said, "The name is Roy Chapman Andrews." Some years later Warner went to hear Andrews lecture on his expedition, and Andrews said, "On this day we met the caravan of Professor Warner from Harvard, and he mistook me for a man named Livingston."

One summer Woollcott invited my parents, my older brother, and myself to his summer place on an island in Lake Bomoseen in Vermont. I remember enjoying it hugely for he had some amusing people there and we played all sorts of grown-up games. Three things stand out in my memory. One is that before our visit Harpo Marx had been a guest.

Some picnickers had decided to land on the back of the island to have their lunch—numerous families with children. Someone wanted to call the police, but Harpo said leave it to him. He took off all his clothes, found a hatchet, and went tearing through the woods in their direction. They were gone in no time. Another was that Woollcott's nice secretary, Mr. Brown, who had a smiling, rather sheepish face and a shock of unruly hair, got in an argument with someone (I forget whom), but I remember the man shouting at him, "You have a face like a loose toilet seat." The third was a play reading one evening. The play was *The Front Page,* an amusing popular Broadway production that took a view of the fast and racy world of a newspaper. My mother had one of the female roles and I read the part of one of the tough reporters, which involved my calling her all sorts of names. Suddenly she stopped the show and said, "You have to give me another part—I can't have my son talk to me like that!" We finally calmed her down.

World events mostly passed me by during those early years of my life, but two things do stand out. One, when I was seven, was the feeling of excitement that infected everyone when Lindbergh had successfully crossed the Atlantic. The other was the great crash of the stock market in 1929. My parents and their friends were playing croquet, and as periodic bulletins came over the radio, I can remember Woollcott shouting that he had just lost his shirt, but he seemed to take it philosophically, almost as a joke.

My father had been working as a vice president in Grandpa's (his father-in-law's) silk business. In 1930 they had a great fight and in a rush we went to live in France. The gloom of our move persisted for some time. We rented an unattractive villa in Vaucresson, outside of Paris, a none too cheery house, decorated with appalling taste. My father was restless and tried with mixed success to occupy his energies writing short stories. My mother was plagued by unpleasant stomach problems and suffered a good deal despite her stoicism.

We boys thought France pretty interesting and did not really notice the drama of our parents' problems, of which we were largely ignorant. We spent a lot of the time with Alphonse and Alphonsine, the couple who did everything in the house. They were irreverent and jolly, and Alphonse was full of boy's tricks that were sometimes cruel but always absorbing. One day he told us he would get rid of a dog that kept coming to the house and pestering us all. He wanted us to be sure we saw the action, which consisted of rubbing the dog's rear end with some gasoline on a rag; the poor dog sped off in horrible agony—it was an image I have never been able to erase from my mind. I had been the numb witness to a torture scene.

I had become an avid stamp collector at the time, and would bicycle to nearby Versailles to squander my allowance in a small, delicious stamp store. Alphonse took us to one-ring circuses and country fairs. We also took pleasant walks in a nearby wood with the unlikely name of Sainte Cucufa where Napoleon's Josephine caught a cold that was literally her death because she wore a wet dress to fully reveal her rather splendid figure. (I don't know if the story is true, but we certainly believed it as we walked around the small lake, hoping she would appear in her boat with her clinging garments.)

Relief came for my parents many months after they had been in Vaucresson. First Alexander Woollcott came for an extended visit, which bucked them up, and then Grandfather Stehli came to make peace. I never knew what was said, but it was successful and my father became more relaxed and my mother slowly came into bloom again.

The three older boys were sent to a boarding school in Switzerland near Coppet, down the lake from Geneva. We were there for three years—I was ten when we began, Paul was twelve, and Henry seven. Despite the fact that it was a period of growing strain for me, and I had begun my general adolescent unhappiness, I loved life in that school. There were about seventy-five boys, from eighteen to seven years of age, yet despite that spread the teaching was exceptionally good. In fact, I skipped a grade automatically when I returned to America, and it was certainly not because of any precocity or scholastic ability on my part; I was a very average student. I do have a few odd memories of the teaching. One was the myopic math teacher whacking the palms of one's hand with a ruler if one had caused a disruption in class, or had made a messy job of copying into one's notebook the daily lesson off the blackboard. Despite the fact that we made great fun of him, we knew as a teacher he was the best, and we loved him for it. I remember our French teacher telling us that Racine was far superior to Shakespeare because he described the same gamut of human emotions using half the number of words!

Now that the great rift had healed, my parents decided that they would continue to live in Europe, but that as much as they liked France, they had to get out of that hideous villa. Because they had many friends in Britain, in 1932 they

BECOMING A BIOLOGIST

moved to London and rented a very pleasant house. Here life blossomed for them both and immediately they entered a whirlwind of social activity. There were two worlds in which they lived: one was made up of their many distinguished American friends who often came to London, and the other was a set of English friends. My father was a strong Anglophile and now he could enjoy that "philia" to its full extent. As an old English friend of mine said, upon reading one of his novels (that he wrote in later life) for the first time, "I never realized your father was such an Edwardian." That was exactly right. All his manners and sympathies were with the British upper class—he considered himself the American counterpart.

I liked all their friends, both the new British ones and the old American ones. One couple, Henry Andrews and his wife, Rebecca West, were particularly kind to us. They talked to us as though we were adults and the compliment was not lost. Henry Andrews had great charm, and I remember with pleasure his taking us to the circus, something he seemed to enjoy as much as we did. His manner was very gentle and soft-spoken, with a smile on his face that was encouraging and sympathetic. He gave one the feeling that this gentleness was no sham, and both a statesman and a child would get the same treatment. Rebecca West gave me a book on birds in Switzerland that I treasured, although, alas, it seemed rather a dull book later.

After the move to England, my parents rented a quaint old house in Surrey for the summer of 1933. The ceilings were so low that the footman had to stoop to serve dinner. (Many years later a good friend and colleague of mine rented such a house while on sabbatical leave and he told me the only way to survive in certain parts of the house would be to wear a football helmet.) One day we were at dinner, for which we had to dress—I have never been able to shake the habit, but now when I come home I put on a pair of blue jeans—and my father asked me where the tennis instructor was, and I said he was "out with his tart." There was a silence, and my father asked me if I knew what "tart" meant. I said I did and repeated the answer of the young instructor, "It's short for sweetheart." That put an end to that bit of conversation but after dinner my father asked me to come to his room. He opened up a dictionary and said to me, in a solemn tone, "Now I'll show you what 'tart' means," and he pointed to a word. I read it and said, "What does 'who—ore' mean?" I no longer remember what came next, but I think he gave up, and my brother Paul filled in the details. It was all well known to me, but spelling has never been my forte.

I had not thought about this story until a short time ago when I reread Evelyn Waugh's *Handful of Dust*. In it there is a very similar story with a small boy and his father. I was quite struck by the coincidence, but then I remembered my mother telling me she had sat next to Evelyn Waugh at a luncheon. She said he was difficult and morose and she had to do most of the talking—I wondered if she had told him my story, which had amused her so much; it would have been at a time when he might well have been working on his book.

For me that period was key in setting a course for my life. While wandering through St. James's Park in London I would pause to watch the ducks under the bridge, and with the help of a set of bird books in our rented house, I worked hard to identify them. My other haunt was the Natural History

Museum in South Kensington, which was within walking distance. The bird gallery had me spellbound. I have an especially sharp memory of the small, exquisite, Victorian glass cages filled with fairy-like humming birds on branches, all posed in different positions. I did my first watercolor of a bird in the museum: it was a duck sitting among the reeds. Subsequently I did a few others but noticed that each one looked slightly less effective than the one before it, which brought me rather rapidly to the end of my career as a bird painter.

All this happened at the same time that I was becoming increasingly involved in my solitary outings in the small wood near our school. It was a narrow strip of trees that bordered on a small brook; for me it was an idyllic spot. It was simultaneously exciting and calming to my emerging self. I could watch the seasons change from fall when the leaves were shed, through the snows of the winter, to the glorious spring—yet all through those wonders the brook would stay the same with minor variations, such as when bits of ice would form along the edge in cold weather. I hardly ever went with friends, of which I had a number; this was a place where I wanted to be alone. In a childish sort of way I felt it was my place.

In the spring, with my mother's racing binoculars and a very primitive camera, I spent hours watching birds and occasionally finding a nest to photograph. The results were very rudimentary, but they gave me extraordinary pleasure. The bedroom I shared with Paul was on the side of the school near the wood, and as we drifted off to sleep in the spring we could hear a nightingale giving its extraordinary full and varied song. Nightingales seemed to me to have the same quality as great opera singers, and I heard them again in 1985 while staying with my brother Tony in Mallorca—my childhood recollection was more accurate than I imagined possible after a span of almost sixty years.

My father worried that I wanted to become an ornithologist and he felt instinctively that this was too narrow an occupation. To lure me on to wider horizons he gave me a copy of *The Science of Life* by H. G. Wells, Julian Huxley, and G. P. Wells, which had been recently published. It was a masterstroke on his part because almost overnight I switched from birds to all living things. The book captivated me and I have always felt grateful to it and to my father's insight. It has a clear and fascinating account of where biology was in the 1930s when I began.

Many years later I got to know Julian Huxley and G. P. Wells. Gyp Wells, a friend who shared my interests in natural history, was H. G.'s son. I could never get him to tell me about the writing of the book, and I do not quite know why, but I suspect it had something to do with the tribulation of dealing with two rather large egos. When I got to know Huxley he had finished all his notable work in embryology, evolutionary biology, and animal behavior and much of his popular writing. He visited Princeton a number of times and I had some easy and enjoyable conversations with him. He was very sensitive about his role in modern biology, partly because he had taken some positions on evolution that seemed a bit mystical to some, including myself. But I greatly admired many of his contributions, such as his work on the development of sponges, on relative growth, and on animal behavior. A story I heard perfectly fits what he was like as a person. He arrived at Yale to give a lecture, and his nephew met him at the train. Huxley was elegantly dressed for his public appearance, and his nephew remarked that he looked very distinguished, to which Sir Julian replied, "I am very distinguished." He was the grandson of Thomas Henry Huxley, who did so much to support the views of Charles Darwin. To close the circle, as a young man H. G. Wells had taken T. H. Huxley's famous course in biology.

Because biology was science, I tried to take a course in chemistry in my Swiss school, but the science instructor felt I was too young. I liked him and it was a great disappointment to me that he had no interest at all in my budding science career. In fact, for some reason he made fun of my reading and my activities in the woods. I could never understand why, although no doubt he found me priggish. It did not deter me; I was just slightly bewildered. Had I been less determined and secure, his effect might have been devastating. At least he taught me how not to behave with one's students. Perhaps his opposition was just the thing that spurred me on. In any event I did not waste much time thinking about it in those early days: I set my own course, which to some degree was isolated from the rest of the world.

The biology taught in school then seems very primitive now. We concentrated on zoology and the textbook was a straightforward description of the various phyla. There was no genetics and no evolution; there were no ideas but almost entirely description. I realize in retrospect that I was getting the real biology of the day from my copy of *The Science of Life*, first published in 1933; while it also offered description, it introduced me to evolution, genetics, embryology, physiology, and behavior. I do not remember noting this then, but I cannot help thinking that, however unconsciously, it started me on a path that led to my interests today. My only direct contact with modern biology of the time was a school trip to Geneva to visit the laboratory of Professor E. Guyénot, who was pursuing some experiments with T. H. Morgan's fruit flies. I saw not only the normal, red-eye flies in a vial with the larvae crawling through a banana mash at the bottom, but also a white-eye mutant, which—I only appreciated many years later—was the first mutant discovered in Morgan's fly room at Columbia University. I think my only thought at the time was how much fun it must be to raise flies and do experiments; I certainly did not have a clue as to the accomplishments of Morgan and Mendel. I do remember that Professor Guyénot was a charming man with a ruddy face and a beret, who clearly enjoyed captivating a flock of young children.

In 1934, after my three years in the Swiss school, my father decided we must return to America for the education of his boys! The reason he gave was that he and his father had gone to Phillips Exeter Academy in New Hampshire and the world would fall apart if his sons did not do the same. That summer my parents lived in the Inn at Exeter with my two younger brothers, and Paul and I went to the summer school so that our European education would mesh with our new American one. I remember the summer vividly because in Switzerland the courses were given in French, and now suddenly everything was taught in English. As a result I did not know the right words for the terms in math (such as squared and cubed), and I pronounced my Latin in a very peculiar way, much to the delight of the other boys. The teachers would make me read the equation, or the passage from

Caesar, as I had been taught, and everyone, except me, thought it was very funny.

During that period school continued to be a heavy burden for I was a poor scholar. My nadir was Latin. I must have spent more years with Caesar's *Gallic Wars* than any other person in the whole history of Latin teaching. It was impossibly difficult and supremely dull. On top of that, my more gifted classmates kept telling me that Cicero was even worse, so that if I ever managed to pass my Caesar, the future would be no better. Toward the end of the year that I was taking Latin II for the second time at Exeter (not to mention the summer school and those times in the Swiss school), my teacher asked me to stay after class. He said, "Do you plan to take Latin III?" I replied, "Not if I can help it!" He then said, "Well, if you promise not to take it I will pass you in Latin II." That shady deal was instantly sealed, followed by joy that has lasted to this day. I have often wondered if all those years of agony might not have had a beneficial effect on my writing (although I prefer to think that if I hadn't gone through all that torture my prose style might have been brilliant!). The only concrete effect was that five years later I received a bachelor of science degree at Harvard rather that a bachelor of arts. The rule then was that third-year Latin was required to become a civilized bachelor of arts.

I was a very mediocre student in school at Exeter in the 1930s. In fact, at one point my middle of the year grades were two D-minuses and two Es, and only one of those could be explained by my nightmare with Latin. At least later I got an A in biology; and I did not fare too badly in chemistry.

I remember my first chemistry teacher with some fondness. He was a thoroughly nice person but awkward and shy.

He would arrive in class in shirtsleeves and wearing a belt and suspenders. Before saying anything, and in front of the whole class, he would give a furtive glance behind him as though to make sure the coast was clear (of what I never knew). Then he would quickly remove his suspenders and stuff them in the top drawer of the chemical bench that served as a lectern. Only after this ceremony did he address the class. How could one not like such a gentle soul? The only thing that upset me was that all during the course I did not see how things held together—chemistry seems to me a rather chaotic natural history of different elements and the compounds they make. It was only at the end of the course that we were told about the periodic table. It struck me like a thunderbolt and everything in the chemical soup fell into place. The elements were related to one another in a beautiful and simple way by the number of electrons that floated around the corresponding number of protons in the nucleus of each atom. Suddenly chemistry made sense, but I was outraged with the old dear for not telling us that in the beginning of the course.

I felt that if biology were taught at Exeter, then I could excel. One morning in chapel our remarkable headmaster, Lewis Perry, announced in his rich, sonorous voice that if any boy had any suggestions for the school, or even any criticisms, they should feel free to come see him. I thought about this for a few days and finally gathered enough courage to make an appointment. With a pounding heart I was ushered into Dr. Perry's office and he beamed at me as he sat me down. I reminded him of his invitation, and his expression suddenly turned black—here was a complainer. I was in too deep to escape so I told him that a school of Exeter's stature should offer biology. The effect was electric and the smile came back double strength. He said he could not agree with me more. He was about to meet with the trustees and he would tell them that a boy had come to urge them to act. I realized later when I thought about it that the move had already been planned, and my timing had been lucky. Years later he would introduce me as the man who brought biology to Exeter—a fiction we both enjoyed.

The next year (my senior year) a young man came to start up the course while he was still finishing his Ph.D. at Harvard. I hovered around Thurlo Thomas as he set things up, and I must have been a tremendous pest. But he was a kind man and overlooked my advanced case of conceit and self-satisfaction. Much to his surprise I got an A in the course, and when he went on to become a professor at Carleton College we sporadically kept in touch. I admired him for his forbearance and forgiveness. The course was my first serious training in biology. It had many good effects on me, the most important of which was the realization that I did not know all of biology, which had certainly been my modest conviction at the beginning of the year. I not only learned many new things, but it was my first exposure to some scientific rigor.

The whole experience had a strong influence on me and I became increasingly excited—my inner conviction that I wanted to become a biologist grew ever more real. As I look back at this feeling, it surprises me in retrospect that one can have such a broad enthusiasm at such an early age. Birds had taken the back seat; now it was all of biology. It was only much later that my interests narrowed down and became more focused. This is the principle of Karl Ernst von Baer, the famous early nineteenth-century embryologist, who, among other things, discovered the mammalian egg. He promulgated his "law" that says that during the development of the animal embryo, first the general characters appear, and later in the life of the embryo one sees that the more specific characters of the species reveal themselves. Well, I was still very early in my development and was quite content with all of biology. In fact, even now I have retained some of this interest in all things biological (despite my eventual special interest in slime molds, of which I will have more to say later). Perhaps it is good that I still have not completely grown up.

It was during this period that I learned to drive. I managed moderately well and finally had to go take my test. In that rural community an elderly man, who looked like a mild form of Teddy Roosevelt, gave the examination. I went to him in his small white house, and told him I was ready to take the road test. He asked me a few questions, the last one of which was did I like classical music or jazz best? After a pause I told him in a somewhat apologetic fashion that I liked both, but classical music was my favorite. He then informed me that I did not need a road test—I could get my license—because in his experience young people who liked classical music drove more carefully and not so fast as those who just liked jazz. I wished he had been my Latin teacher.

Even though I did an average job on the college boards (with the exception of biology where I distinguished myself), my father and some of my teachers thought I was too immature for college and should wait a year. This idea was reinforced when I learned that I had been accepted at Harvard "on trial" because of my general performance. I was absolutely adamant that I would go right away and my determination carried the day. My only object in 1937 was to go to a place where they taught numerous biology courses.

Because I had already taken the basic biology course at Exeter, and taught by a man from the biology department at Harvard, I was allowed to go directly into a second-year course, which in those days was a year of botany. Plants made me just as happy as animals, so I was overjoyed. The first half of the course was on the lower plants and included bacteria, algae, and fungi, and it was taught by Professor William H. Weston.

Cap Weston, as he was known to all his colleagues and all his more advanced students, was a remarkable man who had a great influence over me, probably even in ways that I do not recognize to this day. We all adored him as a person and as someone who understood how science was done. It is quite remarkable that he was held in veneration by so many students, both undergraduates and doctoral students, despite the fact that he himself did very little research on his own. His creativity was expressed through his students: he guided them and prodded them to do clever and ingenious experiments in uncharted waters, and as a result the collective research output from his group was unrivaled in its novelty and originality. And his standards were exceedingly high, so that every aspect of the science and its presentation had to be impeccable. I started doing research with him at the end of my freshman year and continued through to the finish of my Ph.D. This is not the standard way of doing things but an alternative never occurred to me. Part of his wonderful personality included his ability to let his students forge ahead on their own, and he was there to provide the encouragement (or be very blunt if one was doing something silly or wrong). He wanted his students to learn to fly solo, and it never seemed to cross his mind that a good share of the credit should be his.

When I first attended his class I liked him instantly. He was thin, big boned, and over six feet tall, with a mop of very blond hair. He wore gold-rimmed glasses and, even in repose, there was a smile on his face. He wore a hearing aid, a kind that now seems almost as antique as an ear horn. It had a long wire from his ear to a black box that had batteries and a receiver microphone that he had clipped on to his shirt pocket. I have this clear memory of a young woman asking a question in one of the first lectures of the course, some distance from the podium. He smiled, shook his head, and said "hold on a minute." He raced down the isle, detached his black receiver box, and bending his tall frame he placed the box on the desk in front of her, saying, "Would you please repeat that question, my dear."

His lectures were a model. He had a remarkable ability to put elegant sentences together and at the same time manage to make the most mundane topic seem fascinating. He interlaced his lectures with asides and humorous stories that not only kept one in rapt attention, but made it so the biology was hard not to absorb. As a young, eager freshman I had never experienced anything like this, and I was overwhelmed. My first university course in biology not only influenced my future as a teacher but caused me to fall in love with the algae and the fungi, and I decided right away that I wanted to become a "cryptogamic botanist," as anyone working with these lower plants was called at that time. They seemed like such simple organisms yet so varied in their structure and in their life cycles.

The laboratory of the course was a bit of the past that no longer exists but I relished it. It consisted almost entirely of examining various plants, usually live specimens accompanied by prepared slides, and then drawing them with the utmost care. In the beginning one had to learn how to make a scientific drawing. There were set rules: the pencil had to be very hard and sharp, the paper the right kind of drawing board, and the walls of a cell were to have two parallel lines to exactly reflect their thickness. Everyone automatically claimed they could not draw but were firmly told they would learn, which we did. I still have my antique laboratory drawings; they do not show talent but diligence.

Our laboratory instructor was a young and very shy graduate student who was made doubly shy by the new policy, which was to have women from Radcliffe in the course with the Harvard men. Earlier all courses were segregated, and the professors had to give the lectures twice, to the men and women separately. It was still true for some courses, but "Botany I" was a new experiment in coeducation. I nearly fell out of my seat one day when a very fresh young woman who was quizzing the teaching assistant, and who must have enjoyed making him blush, asked in a very loud voice, "Is it true that women masturbate more often than men?" There was an audible gasp around the room and I thought the poor instructor was going to faint. I knew right then the world was changing.

As must be evident, I flourished in Cap Weston's course. It was all I could possibly have wished for. In the middle of it Cap came into the lab one day with my Exeter biology teacher, and I heard them say that, considering, I was doing very well. I'm not sure they didn't mean me to hear—in any event I was greatly excited by the favorable verdict.

My interest in life cycles must have subconsciously started that freshman year, for the variety of life cycles among the organisms in Cap's course was enormous: from simple single-cell organisms that just divide and expand in succession ad infinitum, to huge, elaborate brown algae like kelp and complex fungi such as mushrooms. In the larger forms, it was not just that they developed into something much more elaborate, but each kind of organism had interesting variations in the sequence of its cycle. Cap's course was primarily a sort of comparative natural history: here was the variety of living lower forms and we were looking at them in the spirit of nineteenth-century descriptive biology. It must have stamped a sort of template in my mind, of which I made great use later on.

During the year I had increasingly frequent conversations with Dr. Weston, as I still called him then. It was his suggestion that I might want to take the algae course at Woods Hole that was given every summer at the Marine Biological Laboratory by Professor William Randolph Taylor. I jumped at the chance, but it turned out they only took graduate students, and certainly not a freshman. However, Dr. Weston went to bat for me and persuaded Dr. Taylor I was just the person he wanted. So with my parents' blessing, and enough money to live on and pay my fees, I was off to the Mecca of biologists to take, what seemed to me, the road to bliss. I do not think Dr. Taylor felt he had made a mistake admitting someone so young, although I did one unpopular thing. We had "practical" tests where different unknown algae were put in dishes, and we had to identify them by keying them down in a book. One of the students had lovely curly red hair, and I persuaded her to spare a lock that I put in a dish along with the unknowns. The other students enjoyed the joke, but the Professor was not amused.

Again, this was the sort of biology that barely exists today. One became oriented toward one group of organisms, and one tried to learn everything about them. The first, and classical, approach was to learn the classification of the group and understand their diversity. Once that was under one's belt, then one could worry about the details of their life history, and lastly, which was Cap Weston's strength, one could plan experiments to understand how their life history is controlled.

I was having such a good time in Woods Hole that I tried to seize the opportunity of staying on for the second half of the summer to work on a research project. I asked Dr. Taylor if this would be possible, and to my surprise he agreed and even suggested a problem: to work on the cell structure of a common alga known as sea lettuce, or *Ulva*. He added, however, that I could only begin the problem at Woods Hole, and I would need Dr. Weston's permission to continue with it at Harvard. I was overjoyed and wrote to Dr. Weston. Fortunately I still have his reply. In many ways it was the beginning for me—my first stab at doing laboratory research, and I was thrilled to the bone. Here is part of his letter:

Dear Bonner:

It certainly does my old heart good to hear from you and it is no end rejuvenating to relive in your enthusiastic descriptions my own experiences at Woods Hole years ago. I can see you are getting a lot out of it, both in the work itself, in the contact with interesting workers in other fields and in the association with men like The Admiral (Dr. Taylor to you, my boy) and others.

Now as to the research. So far as I am concerned, the one sentence "I am sure of one thing and that is that I want very much to do it" settles the matter. *Ulva* is an excellent possibility, the alternation of generations which has been worked out in it clamors for further cytological illumination . . . Think nothing of the drawback of carrying on such work here as an undergraduate. That can be arranged . . . I can unofficially assign you a table and equipment in Room 490 and you can carry on with your work as opportunity permits. Nor is the lack of osmic acid too great a draw-back, for I have secured . . . [some] . . . and I am sending it to you by mail . . .

So, with all obstacles more or less cleared up for the time being, plunge into *Ulva* with all good wishes for the best of luck.

Heaven!

During my next year I spent many hours sectioning and staining the *Ulva* I had so carefully collected and prepared at Woods Hole. I was discovering how to make microscope slides the old-fashioned way, embedding the tissues in paraffin and cutting them into thin sections with a special machine holding a very fine knife. It was learning a craft, a skill, that stood me in good stead later on. I loved the process, and I loved the companionship of all the graduate students who would kindly give me helpful advice. I felt the opportunity and privilege that had been bestowed on me. My only disappointment was that the object of all this great effort, *Ulva*, had miserably small chromosomes, and because of the stiff cell walls was hard to section cleanly. The year was not wasted because I learned how to do many useful things. I also learned that it is best to pick one's own problem and not to rely on the judgment of others, no matter how well meaning and respected they might be.

Being a college student in the 1930s was different in many ways from what it is now. In the first place we all wore jackets and neckties for classes and meals. There were waitresses in the dining halls; cafeterias were something that came during and after the Second World War. We also had biddies to clean our rooms and make our beds. I do not want to give the impression that I yearn for those aspects of the old days, because I do not, but I have to confess I do have one regret. Our biddy, who must have been in her fifties, as were all the others, was the sort of kind, motherly soul that all undergraduates need. We confided our minor troubles and anxieties to her and what she gave us in return was sympathetic comfort and good advice. We knew she was on our side, and I remember being especially touched when I was rushing off to a big exam. I said to her, as I dashed out of the room, that I was very concerned about it, but she said, "Don't worry, you'll do well—I burned a candle for you in church last night." I passed the exam and bless her.

In my sophomore year I began to feel as though I was

becoming part of the university. I was now, with my room-mates, in one of the houses and I began taking more advanced courses. For the first time in my life I found myself also enjoying courses other than biology. One was in French literature, given by Professor Allard. His lectures were gems, but he talked as though there was no one in the room—he seemed wrapped up in his stratosphere. This was misleading. Once my brother Henry was visiting and I brought him along to class. In the middle of his lecture Professor Allard stopped, glowered at Henry, and bawled him out severely for not taking notes. There was no chance of my explaining. I loved the reading—all those bits of *les classiques* that I had missed at school.

Some of my other courses were less successful. I seemed to be too greatly influenced by the person teaching it. The subject of geology interested me but the professor managed to dampen all my enthusiasm. In an early lecture he said that a messy desk was the sign of a messy mind and he went on to exhort us to keep both orderly. I found this such a patently absurd bit of sermonizing that from that instant on he was on my black list. He made his second, unforgivable mistake by asking us on our first test what kind of stone were the steps leading to the building made of. I didn't mind getting it wrong, but then to be told afterwards that I had weak powers of observation was a bit too much.

The opening lecture of a course in the history of science put me in such a fury that I instantly dropped the course, which in retrospect was probably a great mistake. It was given by Professor L. J. Henderson, whom later I came to admire for a number of reasons. He did, however, suffer from a protruding ego, and in this opening lecture he said that the study of the history of science was the study of genius, and that resulted in an immediate problem for him. He was able to present the material without difficulty because he was a genius himself; his problem came in trying to explain this to a bunch of ordinary dolts—namely us. I couldn't believe any good could come from that sort of nonsense, but some years later, after I had finished my doctorate, I became trapped by the beautiful prose of the Harvard historian of science, George Sarton, and suddenly saw all the things I had missed, although it was not too late. I wrote to Professor Sarton to complain that I had spent all those years at Harvard and I had taken none of his courses nor even met him. He sent me a charming reply.

My other science courses were a mixed bag. I thought Professor Louis Fieser's huge course in organic chemistry was inspired, not only because of his teaching of chemistry but also because of his skill at handling that multitude of students. His morning greeting was a great hiss from the whole class, which always generated the most wonderful grin as though he were a great concert pianist who had just received a ovation, and we students meant it that way. He did all sorts of things to increase the hiss, such as wearing an incredibly red tie on St. Patrick's Day. One notable moment was when he demonstrated the use of a mordant for a dye. He held up a piece of white cloth, and told us that mordant had been put on only part of the cloth. He then plunged it in a great beaker of red dye making a great fuss stirring it. It came up all bright red. Then he plunged it into another beaker containing plain water, again with dramatic stirring, to wash off the dye where there had been no mordant. Finally he held up the cloth and in great big red letters it said TEST FRIDAY. The hisses (mixed with laughter) reached a new level.

To me the intellectually most satisfying course was in physical chemistry taught by Professor E. Bright Wilson. It stretched me beyond my mathematical capacities, which was an exciting, and frightening, new experience. Many years later I was advising biology students at course-choosing time, and my advisee asked me if she should take physical chemistry. I said by all means. I told her that taking it as an undergraduate it was the best thing I ever did in my biological career. She nodded and asked me with moderate politeness, did I not realize that what I learned in my physical chemistry back then, they now teach in the general chemistry course.

At the end of my sophomore year I had an experience that further helped to guide me on my path. The houses at Harvard were patterned around the Oxford-Cambridge system of colleges, and we were assigned tutors with whom we met periodically. My tutor was a spirited man named Charles Easterday Renn. His claim to fame was his discovery of the cause of the wasting disease of eel grass; it was the result of an infection by a curious amoeba-like organism (*Labyrinthula*), which was of great interest in itself. He was an enthusiast, and he was quite convinced that the most interesting kind of biology was applied, and that is what I should pursue. He suggested I seek a volunteer job in the mosquito-control project of the Tennessee Valley Authority, for he had friends there who could use a young research helper. It seemed to me like an idea well worth trying, so I agreed and went down to the mosquito headquarters at Wilson Dam in June of 1939.

I had never been to the real South before, and therefore

besides the unfamiliar kind of the science, I was to have many new experiences. Everyone was welcoming and kind, and I soon found a room in a boarding house. I remember with affection my motherly landlady who was always full of good advice and help. Her house was in Tuscumbia, Alabama, the home of Helen Keller.

The first shock was the incredible heat—those were days well before air conditioning. I have clear memories of taking a cool shower before bed, lying naked on the sheet, and pouring sweat until I fell asleep.

The second shock was the kind of work we did. We spent days, sometimes up to our waists in the swamps, counting mosquito larvae. Some of the swamps had been sprayed with arsenic, others not, and our job was to determine how effective the spraying had been. I did not mind the out-of-door labor; what oppressed me was the stultifying intellectual scope of the whole project. The small matter of spraying arsenic all over the place only seems grotesque in retrospect. At the time no one foresaw any problems it might create. That was not my difficulty. I simply could not imagine how anyone found such a project of any interest at all. It might have been useful, and indeed malaria was a serious menace, but how could such mundane procedures capture the imagination of anyone? I decided that my friend and sponsor Charlie Renn must be out of his mind.

We also collected adult mosquitoes from houses, and this was my introduction to the very poor, mostly white communities. Their houses were ruinous shacks, their dress and their manner pathetically slovenly. The small children would be wandering about covered with dirt and with sagging, soiled diapers. They talked to us with reserve and distrust. At a Saturday night dance I saw one man in a booth hit another man over the head with a beer bottle and knock him out. An equally drunk, short, fat policeman in a neighboring booth rushed over, and leaning over what looked to me like the corpse, began blowing his whistle nonstop for some minutes. I remember a sign in front of a small village store that said, "Be Respectful to the Women." They were not a happy breed.

My colleagues were a great disappointment, and it was not just that they treated me as the lowest man on the totem pole, which I was. I never had one conversation about biology. Political conversations were as frequent as they were distressing: all the problems of the world were caused by the Catholics and the Jews. Blacks lived in another world that we hardly ever seemed to enter. Once a week all the members of the group and their wives met at someone's house. The hospitality was generous and thoughtful, but after everyone had their beer in hand we would sit around a circle telling mildly dirty stories. The women participated with appropriate blushes and disclaimers. I decided that I was not cut out for applied biology.

In the middle of the summer I got a call from my college roommate. He said his brother had just graduated and was given a car. Could I join them to drive out to the West Coast and back? I never made a decision so fast in my life—I said my good-byes and was on the train to New York in a flash. It was one of my better decisions.

A good way of understanding what modern biology was in the 1930s is to look at the courses I took. There was a tremendous surge of interest in biochemistry, which had the peculiar result that certain current hot subjects were taught in a number of courses, including the biochemistry courses. I think I must have been forced to learn the chemical details of intermediary metabolism four or five times over. The high point for me was a biochemistry course given by George Wald who was probably the best lecturer I have ever known. He also ran the lab himself, but alas the main thing I remember is blowing up a flask containing chlorophyll in an alcoholic solution (I had it attached to a vacuum pump, but it was the wrong kind of flask); the beautiful green solution just covered my lab partner's white lab coat from top to bottom. Fortunately she was not hurt, just green.

Embryology was flourishing, although the progress seems modest compared to what we know today. Biochemical embryology was making an important, although rudimentary, beginning; for instance, the early attempts to identify the chemical nature of Spemann's organizer or inductor were mainly unsuccessful.

By that time population genetics, which was the fusion of classical genetics with evolution, had come into full bloom. It was a fruitful use of mathematics to understand genetic change in a population, and in this way account for evolution in terms of the changes in gene frequencies. The effect of selection, mutation, population size, and other parameters could be given specific values, making it possible to analyze and model evolution in a formal but insightful way. The prime innovators in this important enterprise were R. A. Fisher, Sewall Wright, and J. B. S. Haldane. Once the core ideas had been established, others explored the ramifications for natural populations in what became "the new synthesis," as Julian Huxley dubbed it. There was no appropriate course at Harvard so for one semester Alfred Sturtevant, one of the stars in T. H. Morgan's laboratory, came to teach a course on the subject, which was an exciting event. Our textbook was Th. Dobzhansky's (another veteran of the Morgan fly room) *Genetics and the Origin of Species,* an exceedingly influential book that showed how genetics illuminated evolution.

In later years I got to see and know some of these men, especially Haldane (of which more later). Fisher came to Princeton to give a lecture that I remember well for two reasons: it was on the complex genetics of Rh factors in blood and I did not understand a word, and most of the time he talked he turned sideways to the audience and fixed his eyes on a piece of chalk he kept twirling about two inches from his eyes. One semester in the 1960s Sturtevant came to Princeton to teach a more modern version of the same course, and I had some enjoyable conversations with him; he had a quiet, reserved, almost shy manner that seemed at odds with his inventive lightning brain.

In the middle of my undergraduate years, after I realized that my first research effort was not going to succeed, I became progressively more engrossed in the problems of animal embryology. At first I had no thought of doing research in the area, but the whole question of how an egg turned into an animal with all its fantastic complexity seemed to me an exceptionally interesting matter, one that bristled with questions beckoning for answers. The only point that bothered me was the complexity. I realized that is what I found so appealing about the lower plants: they were relatively simple by comparison. One day the very obvious thought occurred to me—why not have the best of both those worlds? After all, algae and fungi develop too: they started off as a fertilized egg or an asexual spore and grew into an adult. The only difference was that an alga or a fungus adult was simple compared to a frog or a mammal, with their vast number of internal organs and different kinds of cell types. Why not study "embryology," or more properly "development" as it has come to be called, in the lower plants? This seemed to me eminently desirable from a scientific point of view, and it had the key value that I could continue to work under Cap Weston.

Next came the question of what lower organism would be a convenient object of study. I spent hours talking to graduate students and postdoctoral fellows and, of course, Cap Weston. I was seriously tempted by some water molds, a particular passion of Cap's, until one day, quite by chance, I picked Kenneth Raper's Ph.D. thesis off the shelf in the outer office. He had done his graduate work with Cap a few years earlier. It was a splendid bit of work on slime molds that gave me the most wonderful feeling of "this is it."

The cellular slime molds were a little-known group of amoebae that have a very unusual life cycle. They are extremely common organisms in virtually all soils, but they are hard to see, except in the laboratory, because they are so small. As separate amoebae, after clearing an area of bacteria that they eat, they stream together in large numbers to form an instantaneous multicellular organism. This newly created cell mass, in the new species that Raper had discovered, looks like a minute slug, and in all ways acts like a single, multicellular organism: it has a front and a hind end and it will orient toward light and in a heat gradient. After a period of migration it stops, rights itself, and shoots up into the air, forming a minute fruiting body a few millimeters tall in which some of the cells have given rise to the stalk, and each one of the other cells, which are lifted up into the air as the stalk rises, turns into an asexual spore.

I could instantly see these minute beasts had everything I was looking for. Instead of having hundreds of cell types, as we and all other vertebrates have in our bodies, they have two: stalk cells and spores. In the laboratory they are easy to grow, and from a single spore it is possible to get a generation in four to five days. I immediately wrote to Kenneth Raper who kindly sent me cultures and an encouraging note. I felt incredibly lucky.

By the time I graduated I had worked with slime molds for two years. I had not really made any significant discoveries during that time, but it was not time wasted for I made many observations that were helpful to me later. My undergraduate thesis was well received, largely I suspect because I submitted it with a 16 mm film I had made of slime mold development, with the help of the departmental photographer, that was striking. Considering what now seems like antique equipment that I had at my disposal, the good result was especially remarkable. The only horrible moment came after I had spent hours with a small press printing, and then photographing, the title. When it came back after having been processed days later, to my utter distress I had made a simple spelling mistake in large, bold print. Spelling has never been my forte, but this was so bad I recognized the mistake myself. Nowadays with a video camera on the microscope life has become very much easier.

The thesis itself seems to me terribly immature as I look

at it now. Cap Weston was a stickler for writing a paper or a thesis with the utmost care. He felt strongly that it was not only his job to teach one how to do research, but how to present it as well, both on paper and in a lecture. He went over my thesis with his usual care, and when I went to see him, he smiled and shook his head, saying, "John, I hardly know where to begin to tell you what is wrong with this. It is not in a form fit for publication, nor even a standard thesis. Redoing it would be a major task, and anyhow, there is not enough time. The only way I can describe it is to say it is 'pure Bonner.' All you can do is submit it and take your chances." It must now be clear why I think the film helped!

There is a sequel to the story of this immature thesis, the first part of which was on the use of symbolic logic as a tool to analyze slime mold development. In the 1960s I had a visit from an old friend who was a distinguished mathematician, and we were to have lunch together. Just before we went I had some business to do, and I asked him to wait in my office where there were lots of books to keep him occupied for ten minutes. When I returned, to my dismay, I found him reading my senior thesis. I asked him, "With all these good books here, how could you be reading that?" He said it was wonderful, and after I expressed total skepticism, he said, "It's wonderful that you got this out of your system at such an early age."

I see now that there was another important thought that was beginning to form in my mind. As I worked with these slime molds I became increasingly conscious of the fact that their life cycle was quite radically different from those of most organisms. The majority of plants and animals, including ourselves, begin as a single cell that is supplied with food reserves, such as yolk, and after a period of growth and cell division start to feed and bring in energy from the outside world. In plants energy is captured by producing photosynthetic structures such as leaves; in animals once we have formed a mouth and a gut tube we begin to eat. The slime molds, on the other hand, eat first as separate amoebae, and after their plate is clean of their bacterial food, they stream together and develop into a multicellular form, a fruiting body with its stalk cells and spores. They seem to have an inverted cycle where they feed first and then form a multicellular organism, while we do the opposite and grow into a multicellular organism first and then start feeding. More than anything this difference, combined with all those other life histories I had encountered in Cap Weston's course, started me thinking about life cycles and how central they were to the very meaning of life.

The 1930s was an era in which we all felt the gathering storm generated by Mussolini and especially by Hitler. It was a period where the Jewish refugees from the Nazis began to trickle into Europe and America; we all knew it was leading into darkness. I was driving across Texas in the middle of the night with my roommate and his brother when on the car radio we heard Hitler's speech announcing his annexation of Czechoslovakia. It was a chilling moment. Everyone was to feel the repercussions of his evil in the years to come, including his own countrymen.

In 1941, the spring of my senior year, I received a prize fellowship for foreign travel for the coming summer. Had the war not been in serious progress, no doubt I would have gone to Europe but I could not even think of it. I did not realize it at the time but as far as my biological education was concerned, this was a great blessing. I decided to go to Panama instead, stopping off in Cuba on the way back. In my tremendous ignorance I was totally unprepared for the wonders of a tropical rain forest. The experience was such a profound one for me that I feel to this day that all budding biologists, no matter how laboratory-bound, should be sent to the tropics as part of their basic education. It opened my mind—it was a new world.

I had made arrangements to spend most of the summer at the Barro Colorado Research Station in the Canal Zone. When Lake Gatun was created as part of the construction of the Canal a small mountain became an island, which was named Barro Colorado. Because much of it was virgin forest and because it was an island, the Smithsonian Institution and a few universities joined together to turn Barro Colorado into a station for biological research. The location was quite beautiful. The laboratory was built above the boat landing but way above, so there was a hillside of steps between the two. The spacious ground floor was one big laboratory screened in on all sides, which allowed the soft trade winds to waft through.

A rather gruff but thoughtful and kind entomologist named Zetec was in charge and he brought me on the train from Colon, the final lap being across the lake in a small boat. He introduced me to the staff, none of whom could speak English, and I had no Spanish. The next day Dr. Zetec left and I was the only scientist there, and was alone for most of the summer, with the exception of a brief visit of an ecologist. I had never been alone before, and here I was, young and callow, thrust into the life of a hermit in a tropical paradise.

I had come there ostensibly to collect slime molds and related beasts but that immediately went into the background. The stunning, garish birds such as toucans, the fantastic trees and lianas, the army ants and the leaf-cutting ants, the monkeys, the raccoon-like coatis, the peccaries, and on and on, soon became an obsession. I spent all my time trying to identify what I was seeing by consulting the various books in the small library in the laboratory. In my walk each day I would see something new. I spent hours watching the monkeys, especially the howlers and the capuchins. As far as I was concerned the island really was a paradise.

At first it was only paradise during the day. As a child I had always been scared of the dark, and I still was. Nothing had prepared me for tropical night noises. I had a big flashlight and would force myself to go to the beginning of one of the trails, but I would not linger long. The calls were loud, and with the flashlight I would pick out glowing eyes of unidentifiable, scary animals. I was not even safe in the laboratory. Bats flew around inside circling about me, and I had to stun them with an old tennis racket as they flew by. (There was no tennis court on the island—I decided that was the express purpose of the racket.) Once a bat landed right on my chest just as I was getting into bed; it was a new sensation I have no desire to repeat. As I sat reading in the evening occasionally something would hit the outside screen like a baseball. This happened when Dr. Zetec was on one of his weekly visits, and he rushed out, returning with a gigantic, slightly stunned beetle. From then on I collected them for him. A tree frog inhabited the drain of the cement shower. It was a species that had a loud call, but the plumbing gave it a megaphone. When it first let loose I had remarkable palpitations that did not subside quickly. Being alone most of the time did not help, but then perhaps it did, for slowly the fear receded, and from that summer on the night seemed tame.

I had to do most of my learning without a teacher. I spent the evenings reading Proust, which was perhaps a unique way to learn tropical ecology. For a few days there was a visitor—a distinguished ecologist of an earlier generation and he taught me some things, but we spent more time arguing for he believed that the important thing about nature was its complexity, and we could learn nothing from experiments; that was interfering with nature. I remember the high point came when he told me that all of genetics was bosh because it was done with flies in milk bottles, which reminded him that in his youth he had earned money as a milkman! I was quite polite about it all, repressing a seething, youthful outrage.

I got along very well with the staff despite an almost complete language barrier. There was a breadfruit tree just outside, and I wanted to try eating it. So I asked Rosa, the cook, if I could have some "pan frito." She gave me an odd look, and that night I got two slices of plain bread that had been fried. I went to get her and took her out to the tree, and she said, "Ah, fruta de pan," and laughed. Then she explained with great care that it was no good unripe, and when ripe the animals get it. So I still haven't had breadfruit, but I'm told I haven't missed much. I also took my laundry to Rosa the first week and made all the needed gestures. She said "si, si" and rushed off to bring me a bar of soap and a rock, pointing to the lake, down those thousand steps. I decided that was a skill I was not in a hurry to acquire, so I would lay my clothes out on the grass just before the daily drenching cloudburst, then let the sun dry them, and retrieve them before the next rain. It worked to a point, but I did smell somewhat mildewed.

Later in the summer I went to capitalist Cuba for a few weeks. Even though my headquarters was the Harvard Botanical Garden in Cienfuegos, I spent a good deal of time in Havana going to jai alai matches, drinking beer, smoking Romeo & Julietta cigars, eating delicious sandwiches at Sloppy Joe's bar, and sewing wild oats. I have not been back to Cuba since then and have fond memories of the special atmosphere of the place. In Havana the streets and the buildings seemed quaint and Latin, the people friendly and attractive; everything together made it a delightful place for a young man to roam. I remember visiting the state capitol, but they would not let me in because I did not have a jacket. For a small fee, however, I could rent one. It was all good medicine for an ex-hermit.

This twenty-year era, 1940–1960, was one of major change for the whole world, for biology, and for myself. It began with the Second World War, which had already started in Europe and we and the Japanese joined in 1941. Almost everyone in the world was affected in some way, and for many with the worst kind of tragedy. During the war years biology was rather quiescent, especially compared to physics, but it began to produce huge blossoms after the war.

When I was in my late teens, during my last years in school and my early years in college I had my first great love. It was a fundamental experience for me, undoubtedly because of its novelty. Nothing in my previous life had taught me to expect that one could feel with such intensity for another person. It hit me like a tidal wave. I realize now that a combination of genuinely liking a person and all those adolescent sex hormones is quite overwhelming. I thought no one in the world could be as excited and as happy as I was, and that my situation was unique. The young woman, the daughter of one of my teachers, was quite beautiful and intelligent. What I could not understand was why she reciprocated my feelings—how could anyone be that lucky? I was intoxicated, and I pretty much remained in that enviable position for over three years. She was also interested in biology and we had many things to talk about, but after all these years I can no longer remember what they were.

What I can remember is touching one another. By modern standards of youthful behavior we were circumspect, yet curiously it was very satisfying. I do not remember being frustrated because we never bedded; I recall only pleasure in all those delightful but rather elementary things we did. Youth has many disadvantages, and I never minded when the mature years crept over me, yet a first love is something quite perfect and a wonderful stepping-stone to maturity. I have always felt guilty that it did not end gracefully for I know I handled it badly. Finishing is much harder than beginning, with all its euphoria. No doubt we grew apart from one another, which is hardly surprising at such a young age, but that made the disengagement no less easy.

My mother was quite aware of the affair from the beginning to the end. She was totally noncommittal about the virtues of the young woman. She took no sides—she seemed to look upon the whole thing as something uncertain, no doubt because we were so young. At the time I resented her attitude a bit, but in retrospect it takes on the aura of another example of her wisdom, for indeed the passion ran its course.

Later I met a young woman also majoring in biology who seemed particularly attractive. She was taking comparative anatomy and I would go down and chat with her in the anatomy laboratory. My thoughts did not really progress but we remained friends, and I remember the moment when I suddenly knew that was the way it had to be. I was walking up the staircase and I suddenly got rather a strong whiff of formaldehyde, and like Pavlov's dog I immediately saw her face in my mind's eye: clearly there were some problems with this relationship.

It was not until much later that I met someone who from the very beginning had a profound effect on me. I went to a party in Cambridge and sat next to a young woman full of animation and quick wit; we immediately clicked—and to top it all, she had glorious red hair! She was a senior and I was a first-year graduate student, so we could carry on a blissful courtship far removed from any parental scrutiny. This went on with increasing intensity and joy for months. In 1941, as we were walking side by side in the mist down a Cambridge street, I proposed and was accepted. A happy moment.

The next step was telling our parents. We made a visit to Ruth's mother and father in New Jersey, where the anticipatory anxiety was far worse than the event. I had a splitting headache in the railway station on the way down, and I tried to swallow some aspirins without the help of a drink of water—a mistake I have never repeated. Ruth's father was a wise man, and when I was left alone with him he produced a bottle of whiskey and kept up the small talk until we had made an appreciable dent in the bottle. When he finally paused it was easy for me to ask him if I could marry his daughter, and easy for him to say "yes." It was the beginning of my great admiration for him.

Our visit to New Hampshire to see my parents seemed an even bigger mountain to scale. We went there for the weekend bringing an old friend who, poor fellow, was supposed to buffer the explosion, should there be one. My mother had, of course, figured out exactly why we had come, but neither she nor my father had ever met Ruth. Ma often told us that if we wanted to know if our friends were suitable, we should bring them home and then we would immediately know. We were about to face the acid test, bringing our diplomatic friend along, whom we knew they both liked. We had not been there very long when my mother took me to one side to "have a little talk." Without preliminaries she said, "I like her very much and it is obvious you plan to marry and that is why you have come here. You have my approval, but I know you well enough to know that you will marry with or without it or that of your father. There is only one thing you have to do. Don't tell your father—ask for his permission." I felt as though this would be the last time in western civilization that a son asked his father such a question. I even had a fellowship that was sufficient to support us—but of course I obeyed. Ma could phrase such a request in a way that left no option. When I asked him he was pleased with the decorum of the question but even more pleased with the thought of his prospective daughter-in-law, whom he liked from the very beginning. Because of her I rose in his estimation.

A bit later Ma had a very nice conversation with Ruth in which she told her how pleased she was. She also felt she should forewarn her daughter-to-be that all would not be a bed of roses—while I was interesting, I was also very pig-headed! The nice part is that in later years, in her usual way of saying just what was on her mind, Ma would periodically tell me that I had found exactly the right person.

After a few months of marriage, Ruth and I visited Dayton, Ohio and I gave a lecture on my graduate student research. The lecture was arranged by Charles Thomas, the charismatic head of the Monsanto Chemical Company, and after the lecture an Air Force colonel came up to me and introduced himself. He was the head of the Aero Medical Laboratory at Wright Field, and they were looking for young people like me to do research. What were my army plans? I said I had none and was about to be drafted. He told me to volunteer and then gave me detailed instructions on where and how so that I could be transferred to his laboratory.

These were carefully followed and soon I found myself at an induction center having a physical. My first impression of the army was very strange: an endless line of naked men, every one with a folder of papers in his hand, going from one room to the next to have every part of his body checked and his chart filled in. The only interview with any privacy was with the psychiatrist. He looked at my chart and then asked, "Do you consider yourself normal?" Without too much thought I said "yes." He jotted something down on my chart and called for the next person. As soon as I emerged I could not resist the temptation to look at what he had written. It said "officer material"—I suddenly feared for our success in the war!

Ruth went with me to the army base where I was to begin my army career. It was not a happy day—it was a parting into uncertainty. The idea that we might soon be together again did not cross our minds.

At the base I was processed. A complete uniform and all the duffel I would need. The part I liked best was the ancient sergeant who wrapped his hands around your neck as if to choke you, and then yelled your collar size to a private who handed out the shirts. I always thought they could use him at Brooks Brothers—it would liven up the place. I went into a large room with many desks and was told that I was to be classified. A burly sergeant asked me what I did. I told him that I was a biologist. He began looking it up in the big book in front of him, and said there was no such occupation listed in the register. I said I was sorry to hear it because that was my only occupation. He asked me to wait a moment and he returned with another large book, and finally found it. Then he asked, "Are you skilled or unskilled?" I said they did not rate biologists that way. He then asked how long had I been doing my work. After some quick calculations I said, "About six years." He replied, "Christ, if you're not skilled after six years, you must be no damned good." So I entered the army as a skilled biologist!

My life in the barracks at Wright Field outside of Dayton

was remarkably pleasant. The majority of the men were from central Pennsylvania, some of them coal miners. They were fine people, but when they got mad at one another the insult was always "You goddam coal miner." The best of them all was Master Sergeant Mertz who had total authority over the barracks and was liked by all. They realized there was something peculiar about me but they were tolerant, and even amiable, provided I rode the waves of their rather crude humor when it was directed at me. Despite all the rough talk, whenever I got in any kind of difficulty they were right beside me to help. They taught me many things that kept me in good stead in later life.

My first job at the Aero Medical Laboratory was to mop out the colonel's office. I remember carefully parking my mop outside the door before knocking—I thought I could introduce myself more effectively without holding a mop in my hand. Fortunately he was not in, but later when I met him in the hallway and said my bit, he seemed totally uninterested—I was not the bright young man he had recruited but the menial help. The idea of research was clearly out of the question. My lot was barracks life, KP (another educational experience), and working in the high-altitude chambers in the basement of the Laboratory. These were large cylinders the size of airplane cabins with benches inside. They could be evacuated by a pump so that one could simulate any altitude from ground level to well over forty thousand feet. We were the guinea pigs for testing many different kinds of oxygen equipment, especially masks. If one stayed up at a high altitude for any period, one suffered aches and pains in one's joints. These are the "bends" due to bubbles of nitrogen gas that form in one's blood. You could feel the bubbles along the vein in the arm—it felt like having Rice Crispies in one's circulatory system. Bends are easily cured by coming down to lower simulated altitudes. The other problem was having clogged ears when going down too rapidly, a familiar difficulty when descending in an airplane. Some of us had large Eustachian tubes, which connect the middle ear to the mouth, and were considered useful guinea pigs because we could be brought down fast.

There were a small number of us graduate students, all of whom were quite cheerfully doing these manual things, but at the same time wondering why we had been recruited by the colonel, who seemed to be unaware of our existence. It turned out that we were not forgotten. We were all rounded up and urged to volunteer for Officer Candidate School (OCS) for a three-month training course on how to become an officer and a gentleman. I agreed, but I nearly did not make it because the day before we were to go, our captain, who was our officer boss, called me in to tell me that because of my French he had just received orders that I was to be transferred to intelligence (and no doubt spend the duration reading other people's letters). Captain Murphy was a splendid fellow who was well liked by the enlisted men. Part of our admiration of him came from the fact that his previous career had been in burlesque. He was the good-looking straight man who introduced the different acts and sang sugary songs in between. He asked me what I wanted to do, and I said I wanted to go to OCS and come back to the Aero Medical Laboratory, for I knew then I could do research. He said fine—he would put the other orders in the back of his desk and not find them until after I was gone.

OCS was a remarkable experience for me. To come under

such total regimented discipline night and day had a most curious effect on my psyche. I felt as though I were changing into a mindless automaton. Shortly after I returned to Wright Field I read a book by someone who had suffered in a German concentration camp, and to my amazement it exactly described my mental separation from the self I knew. It seemed to me that I not only did everything without thought when told to do it, but I seemed to enjoy the automatic nature of the process. Even stranger, I began to like my bosses, not because they were nice men (which no doubt some of them were) but because I felt powerless to think anything else. Ruth sensed some of these things from my letters and from me when I returned, but I never wanted to discuss them with her or anybody. Everything was too odd to talk about. There was no permanent damage, but I will never forget the strange experience of following the actions of someone else who lay within me. I was two people in one, and although it was not too apparent to me at the time, I understood it more clearly later.

The experience had a perverse effect on me. Many of my fellow inmates told me later that my cheer and constant joking about the absurd side of things made life much more bearable for them. I do not remember much of this—again, I expect it was automatic. I remember one bit of sabotage that gives me pleasure thinking about even today. On our very first visit to the athletic field we were asked to do as many push-ups as we could and to give the number to an officer who recorded it. It dawned on me, as the weeks slowly moved on and we spent so much time doing fitness exercises, that they were going to give us the same push-up test at the end, so they could congratulate themselves that they had not only improved our minds but our bodies as well. I explained this to all my companions and told them to pass the word, and especially to be sure to do fewer push-ups when we were tested just before we graduated. The idea spread like wildfire, and we cherished the thought that perhaps the brass would conclude that they tore us down rather than built us up.

What reinforced the unreality of the whole period was that all this took place in lush hotels in Miami Beach that had been commandeered by the army. We lived all jammed up in fancy rooms with fancy bathrooms; we ran on the beach among all the glitter. We were in a world that did not exist. We ate in a big ballroom that had been stripped and refurbished in plastic. I can still hear the tunes they played on the jukebox while we ate. And our table manners were supervised; we had to learn how to eat—one hand on the table and the other on the lap. This presented special problems with large potatoes, but fortunately it was quite acceptable to spear the whole potato with a fork and gnaw at it as though it was an ice cream cone—as long as one hand was in the lap. Becoming a gentleman took on new dimensions.

I enjoyed the classes because for the first time in my life I could get good grades without working. I even spent some time coaching others who were having difficulties. The subjects were stultifying: army law, leadership, supply, and similar practical subjects. The short-answer tests were especially puzzling because they seemed to have little to do with the subject matter; often the questions were impenetrably unclear, and those impossible questions would appear more than once on the same test. I finally screwed up my courage and asked the young lieutenant who taught the supply course about all these peculiarities. He laughed and said the tests were prepared by a special unit of educational psychologists. What they did is cull every two weeks those answers that the bright students answered correctly and the dull students incorrectly. This fortnightly unnatural selection, in no time at all, produced those incomprehensible and irrelevant questions.

When we returned to Wright Field in our brand-new uniforms and shiny gold bars, looking tanned and thin, we immediately began to do interesting high-altitude physiology. There were some excellent scientists in charge, and we quickly became involved in all sorts of urgent problems facing aviators in combat. It took a bit of adjustment, and at my first meeting to discuss a project I had great difficulty following what was said. I was sure OCS had ruined my mind permanently.

We had to report daily to Washington on our research progress. At first this seemed an impossible task: how could one do something worth reporting every day? The unacceptable alternative would be to say that nothing worked that day. I devised a special strategy. If something did work, stretch out the success for at least a week of reports. It gave them suspense—like a serial. I got so good at it that I began helping others with their reports, giving lessons on the rudiments of daily report writing. Once I had to write a larger report summarizing a whole series of experiments. My colonel, a fine man who was sometimes a bit irascible, being a professor of physiology at a medical school and chafing at army ways, sent the report back with a scribble in the margin saying "rewrite." I was incensed for I thought the report rather good. I asked a friendly secretary if she would enter into a small conspiracy with me: she retyped the report without changing a word on a typewriter with a different font. I handed it back and received a courteous, professorial note thanking me for the vastly improved version.

After the slow and frustrating wait to get discharged from the army I returned to Harvard to finish my graduate studies. The largely descriptive work done during the year before I left for Wright Field was published during the war with much help from Cap. I still have our correspondence about the paper, and Cap shows a characteristic concern that I would be upset by the referees' reports. He gave me a lesson by correspondence on how much the first criticisms of anonymous referees could cut to the quick, and how one must push aside the harsh way the comments are clothed and pay attention instead to what they are suggesting, for their advice in this case was worth following. His counsel was a great help when he finally sent the referees' comments on to me; it has protected me and many of my students from those often brutal blows. The protection is never total, and especially in early days one is vulnerable.

I now had to produce a thesis, and I knew that for the first time I had to discover something new—I had to go beyond description. The thought did not make me anxious because I was so deeply involved in the excitement of day-to-day research, which seemed to put me into a trance-like state. Because slime molds develop so rapidly I was able to grow them so that I always had some culture dishes ready at each stage of their life history. I still have those notebooks, which are small volumes with cramped writing and many sketches of what I saw. That great accumulation of trying different things has paid off in many ways; I noticed odd results that I am still exploring to this day. Much more important at that time was that my attempts led directly to the results that gave me a thesis.

I decided during the course of this "playing" in the laboratory that my main object was to find out the mechanism of the aggregation of the starved amoebae. It could be seen especially well in the time-lapse film that at a sudden signal they all elongated and streamed into central collection points that became the migrating slugs. In retrospect, now that we know the answer, the problem seems so simple one wonders how this could have been considered a major puzzle. But at the time the idea that cells in development could be attracted to one another by a chemical attraction, or chemotaxis, was very much frowned upon. This was because a distinguished and forceful embryologist, Paul Weiss, had made some important discoveries with animal embryos where he showed that chemotaxis did not play a role in some of the organized cell movements in the embryo; the cells felt their way along the texture of the surface by what he called "contact guidance." Later I got to know and like Weiss and he gave me some useful advice, but he always cautioned me not to be too complacent with the thought that chemotaxis might be responsible for aggregation in slime molds.

I tried hard to keep an open mind and investigated the possibility that the amoebae were oriented by some sort of electrical force or by some interfacial phenomenon happening on the surface between the moving amoebae and the substratum in an attempt to demonstrate Weiss's contact guidance. I even tested the idea that a center might be giving off some sort of ray because in those days people were still talking, although with considerable skepticism, about "mitogenic rays" that were supposed to stimulate growth. None of these things worked; all my experiments seemed to rule them out. At the same time I could not prove chemical attraction either.

During the course of this work I had developed a way to have aggregation occur on the bottom of a glass dish under a layer of water. One day, to see if a current affected the orientation of the amoebae, I decided to swirl the water very slowly in a circular dish with a bent stirring rod over some aggregates. I left the motor running and after some time glanced through the dissecting scope to see what happened. I was really not expecting much, and what I saw nearly blew me through the roof. The current had produced an asymmetrical aggregation pattern: there were no oriented amoebae upstream of the center; they seemed to be wandering about aimlessly, while the amoebae downstream were perfectly oriented toward the center and moving against the current. In a flash I realized the attraction had to be by diffusion, and the diffusing agent had been moved downstream by the current, like wind moving the smoke from a pile of leaves in the fall, with no smoke upwind, and the smoke trailing long distances downwind. What surprised me afterwards was how quickly I read the message sent to me through that one glance into the dissecting microscope. Instantly I saw that chemotaxis had been proved, and that I had made the discovery that would get me a satisfactory thesis. I remember dancing about my lab room and punching the air in my excitement.

The experience also taught me a great lesson. I had not carefully designed an experiment that would prove diffusion;

I had managed it by accident. That and all the other observations I had made told me that the slime molds were in charge, not I. They would let me know their secrets on their terms, not mine. A gifted and delightfully eccentric mathematician who helped me with the publication of these results knew the same thing: he would write an equation, stare at it for a bit, and then as though I were not in the room, he would say to the equation, "speak to me, speak to me." Well, the slime molds had spoken to me.

In writing up my discovery that the aggregation of these social amoebae was indeed by a chemotactic attraction to an unknown chemical given off in the center of the aggregate, I realized I had to give the chemical a name. The group of which these slime molds are members are the "Acrasiales," and in Edmund Spenser's *Faerie Queene* there is a witch named Acrasia who attracted men and transformed them into beasts. This seemed perfect for me because the chemical attracted the amoebae and they were transformed into stalk cells and spores. So I named the attractant "acrasin."

As I look back at this early success I realize that it epitomizes my approach to experimental biology. First one does the biological experiment to reveal a new phenomenon and then that discovery can be exploited by further, largely biochemical experimentation: in this case that meant finding out what the chemical is and the details of how it orients the cells. Later I will discuss some of the discoveries that followed; let me simply say here that this early demonstration of chemotaxis has led to a large amount of work in many laboratories, including my own.

Cap Weston was, as usual, full of encouraging support and I turned on the heat writing the thesis. This time there was

no nonsense, no "pure Bonner"; it had to be Cap Weston perfect. How I slaved, and how he slaved over me. The final
result was short, but the main portion almost ready for publication. My final oral exam was a complete letdown (as had
been my oral general exam before the war). The reason
undoubtedly was an excess of adrenalin and shame at answer-
ing some questions so stupidly. In this final I felt none of the
professors present, including my dear Cap, had any appreciation of what I had done, and that all they did was carp. None
of that, of course, was the case, but the excess of emotion
made it an exceedingly warping experience for me.

In looking back at that period I am greatly struck by the difference between what we know now and what we knew
then. Biology has taken such gigantic steps forward that I
feel as though I was a graduate student at least a century ago.
When I was a student, in many ways we were still in the
nineteenth century, for the main ideas, the main problems,
that preoccupied us had already been spread out in the latter
half of the last century and the early part of this century. In
the study of how animals and plants develop, the main
framework had all been laid down mainly by the great German embryologists and botanists in the 1800s. The new
and exciting way of looking at things stemmed from Spemann and Mangold's discovery of the organizer in the newt
embryo that I mentioned previously. It led to the realization
that there was a chemical communication system and that
development was governed by a series of well-organized signals and responses, which were somehow so well orches-
trated that ultimately a consistent adult was produced each
generation. The idea that development involves such a

sequence is an idea that still holds sway. The only difference between now and then is that now we are acquiring a clear idea of what these signals are and how they are received. We now are able to uncover the nature of the chemicals that do the signaling and the receiving, something that was mostly out of the question when I was a student. I can even remember saying that the identity of the chemical signal was just a detail—it was the principle that was important—words that now seem not only naive but remarkably shortsighted.

The great difficulty for me was that all this splendid early work was written in German, and none of the classic papers were less than eighty pages. (At an early stage I asked my Swiss grandmother to help me translate a paper on sexuality in a unicellular alga. At first she became quite angry—she wanted to know what sort of degenerate things were they teaching nowadays—but I managed to persuade her that the subject was quite harmless.) I thought my struggles with Latin were bad, but German nearly did me in. I not only had to read those papers but present their results in front of fellow students in graduate courses. Worse, I could not get my doctorate degree unless I passed a German exam. How I slaved! I took courses in basic German, and they were agony. I could not even remember words, let alone have a clue on how to untangle a sentence. I had a vocabulary vest with four pockets, each one filled with cards with the German on one side and the English translation on the other. One pocket contained the words I almost remembered, another the words I never seemed to be able to master, and the other two pockets were for words that were somewhere between these two extremes. All day long, at meals, in the lab, everywhere, I would keep flashing the cards and making pitifully slow progress. I was sure I would never get my degree, but I did finally pass the test. Soon after I went into the army and never saw a German scientific paper for four years; by the time I got back to developmental biology I had completely forgotten all my German except for the opening sentence of Genesis in the Bible, which was not terribly useful. One good thing for me came out of that horrible war—English became the universal language of science, and I did not have to start learning German all over again.

Even though I took all the biochemistry courses that were offered at the time, I always found myself pulled to the biological questions. This is something that has stayed with me, although from the beginning I appreciated the great importance of biochemistry, and later molecular biology, to my own work and to developmental biology in general. In my research I often tried to link the biological with the biochemical, but always did so by collaborating with someone clever who could fill in for my deficiency. This led to many fruitful joint projects that have been especially enjoyable because of the collaboration. At the same time I never regretted that I kept my feet firmly on the biological side. There are a number of reasons for this. In the first place, that was the right level for framing the key questions—the problems I wanted to solve were biological problems, and if the answer involved molecules, so much the better, but one must never forget the biological question. Another reason was that I was interested in all of biology, and not just development, and that included genetics, physiology, behavior, ecology, and above all evolution. I wanted to be near all those to see how they interconnected as well as how they were based ultimately on molecules.

None of this could have passed through my head in my student days—I groped forward by some sort of mindless instinct. I did know that biochemistry was at that time the coming fashion, and to some degree I sidestepped it because I had no intention of running with the pack; I was beset by my desire to be independent. All these are more or less rational reasons for remaining a biologist first and treating biochemistry and later molecular biology as secondary. No doubt the real reason is that somehow biology was my way of thinking and I unconsciously followed that path. There is no question that there is power in this kind of twentieth-century biology, and the whole acrasin story and how it has developed over the years are a good example. Many years later I put in a grant proposal to the National Science Foundation and was turned down. I happened to know someone there who was in the NSF administration and asked her if she had a clue as to what went wrong. She said she would have a look and called back the next day to say that my foray into biochemistry seemed awkward and uncomfortable; please rewrite the grant stressing the biological problem where they knew I would be on solid ground. I did and was awarded the grant, showing that others knew me better than I did myself.

The 1950s saw enormous upheavals in the progress in biology. Genetics first became molecular when Delbrück, Luria, Benzer, and others decided to use viruses instead of pea plants or fruit flies to do genetic experiments. Viruses are not burdened with a cell and its complex metabolic machinery—they are parasites of other cells, and therefore they consist largely of pure DNA with a few associated proteins. They multiply very rapidly and produce large numbers of progeny, and taking advantage of these properties it was possible to do genetics in a fine-grain way on DNA itself. It was in 1953 that James Watson and Francis Crick made their great discovery of the structure of DNA that was the birth of molecular biology. For the first time it was understood how DNA replicated itself, and from there much ingenious work was done to show exactly how a particular stretch of DNA (a gene) was able to go on to produce a particular protein. This was the beginning of what has become a vast and important scientific enterprise that has had its effect on all the corners of biology. The most recent triumph has been the identification of the entire human genome. It has given us a handle on how to dissect the steps in the development of all organisms, plant and animal; it has been possible to probe inside the cell so that we know the genes for many of the cell components; it is now possible to examine the relatedness between organisms and build ancestral trees; genes for all sorts of behaviors have come to light. Note that in this list we see that molecular biology has become an important way to study many aspects of developmental biology, cell biology, evolution, and behavior.

One way to comprehend this explosion is in my own field. Since the birth of molecular biology in the 1950s the number of people working on cellular slime molds has gone from two, when I started, to well into the hundreds, and the number of publications per annum has gone from two or three to about one hundred and fifty, and most of them are on the molecular aspects of slime mold development. This is just a view of a small corner of what has happened in all of biology; the number of biologists in general and the number of new journals have been increasing logarithmically since the 1950s.

The changes in biology that have occurred since then not only involve advances in ideas and great new discoveries but also a fantastic eruption of new techniques and new miraculous instruments for one's science. My graduate student days preceded electron microscopes or the many new types of optical microscopes; cell counters and cell sorters; a myriad of biochemical devices such as all the types of chromatography to separate mixtures of molecules; gas chromatography to identify the chemicals in the mixture; and more recently the miraculous machinery that goes with molecular biology; I could go on and on. As a student one summer after the war at the Marine Biological Laboratory at Woods Hole I can remember doing many experiments on slime molds with nothing but basic glassware, a kitchen pressure cooker to sterilize my culture dishes and media, and a hand-operated centrifuge to wash my amoebae free of bacteria. And I thought I had everything!

Perhaps the most notable technical innovation that occurred during this period was the invention of the electron microscope. One could suddenly see the parts of a cell at a much higher magnification than had been possible. The development of the machine and its use were not instantaneous but progressed slowly. One of the early machines was developed by RCA, and James Hillier in RCA's Princeton Laboratories pioneered in perfecting it so that it could be used effectively as a biological tool. He asked me if I would like to use his methods for fixing and making thin sections of my slime mold amoebae. The result was by later standards incredibly crude and the photographs that resulted bordered on the useless, but soon after the techniques were perfected by many laboratories, the emerging results were stunning. Parts of the cell not known to exist suddenly became visible; a whole new microworld opened up. This was followed by an era when the biochemistry of the cell components was linked to the newly discovered cell structures; all in all the electron microscope became an explosive advance.

There were other leaps forward in biology during that period. Animal behavior was going through a revolution. It was a venerable subject, but it suddenly took on a radical new life with the ideas and the wonderfully ingenious experiments of Konrad Lorenz, Niko Tinbergen, and Karl von Frisch. They showed that animals have specific responses to specific stimuli (just as in development) and they were able to make the notion of "instinct" respectable and acceptable; it had been banned from our vocabulary when I was a student. There are innate responses and learned responses. This not only led to advances in our understanding of behavior, but made it possible to ask genetic questions about behavior another subject that always flirted at the borderline of a taboo. Furthermore, animals, even bees, had remarkably complex behavior, as von Frisch made clear in his beautiful experiments showing how scout bees could tell the other bees in the hive the direction and the distance of a new source of nectar.

Konrad Lorenz, the pioneer who showed how birds could be imprinted at an early age to follow their parents, came to Princeton to give a lecture some years ago, and he was a wonderful showman. His lecture was without doubt a marvel, full of bird calls and bird postures, along with a wonderful grasp of the mood of his audience. The next day a colleague and I took him to our "perception center" in the psychology department where there was a series of rooms, each of which illustrated an optical illusion. For instance, with one eye one could peep through a hole, and if two people of equal size were in two corners of the room, one seemed a giant and the other a dwarf. When we arrived there, it turned out that Niels Bohr, the famous physicist, was going to make the tour at the same time. We were all introduced, and Bohr and Lorenz were like two excited children, each looking through the peepholes or standing in corners, having the most wonderful time. After we finished, Bohr asked if we could have some coffee and discuss what we had seen. This was enthusiastically seconded, and when we sat around in a circle, Bohr began to talk. I should say that Bohr mumbled terribly in a thick Danish accent—he was exceedingly hard to follow. But we struggled, and Lorenz, a brilliant talker, kept trying to say something, but Bohr ignored him totally and went on serenely with his "discussion." I have never seen such frustration—poor Lorenz was beside himself. After a half hour it was over and none of us got a word in edgewise. It was fascinating to see the two egos clash, each with his own technique of dominating a conversation. In this case it was a knockout by Bohr. Later I tried to put together in my own mind what Bohr had said in his monologue, and it dawned on me that his message was: things are not always what they seem.

Another way I saw biology change came from teaching. Off and on for forty years I taught the elementary biology course at Princeton. Not only did I have to expand to new subjects that did not exist before, but often new discoveries meant that what I had been saying was no longer true. That still bothers me and I was sorry I did not keep track of the old falsehoods and new truths, so that I could send old students bulletins of all the misinformation I had given them.

A great moment in my life came when as a finishing graduate student I was asked to give a seminar at Yale. I assumed it would be a small group interested in development and was struck dumb with terror when I got there to find out that it was to be before the entire botany and zoology departments in what seemed to me a gigantic auditorium. Fortunately I had brought along the slime mold film that had been the mainstay of my senior thesis, and I was able to talk about my experiments that seemed to prove that the amoebae aggregated to central collection points because of a chemical attractant. I managed to get through the lecture without fainting and was again struck, as I had been before, by the questions. There was relatively little asked about my chemotaxis experiments but everyone wanted to know about the organism itself, which had such a peculiar life cycle and which was generally unknown at the time (while today it is in every elementary textbook). In thinking back about this I realize that unwittingly I had infected the audience with my own early, and almost unconscious, fascination with life cycles.

Afterwards tea was served, which I desperately needed, and as I was sipping it with a shaking hand, old Professor Ross Harrison, long retired and greatly admired as being the most profound American embryologist, came up to me and said, in his gentle voice, "Bonner, if I could start all over again, I think I would work with slime molds." Never has a young man received such a boost as I did at that moment.

I finished my doctorate in the spring of 1947. By then I had been under the wing of Cap Weston off and on for six years— I was now his finished product. He followed my progress in later years with fatherly pride. Yet as I look back I must have been a very difficult and prickly child. It is hard to strike out on one's own and shake the habit of leaning on parental support. I remember his asking me that last year why I did not consult him more, and my replying that I felt about him as I felt about my parents: there was no end of respect (and if I had been older, I would have added love) but I wanted to fend for myself; I wanted to feel independent.

Yet with all that commendable desire to be my own man, I know that there is a large amount of Cap in me. The way I lecture, the way I write, the way I tell stories. Perhaps I would have been that way in any event but somehow I doubt it. When I was a beginning assistant professor at Princeton, I was going to the parking lot to get the car when I ran into an admired friend from the psychology department. He asked how things were going, and I had to confess I was a bit down: I had just come from a biology faculty meeting, and we had wasted an hour discussing why one of the graduate students was failing. Everyone had a theory, one more absurd than the next. One colleague had even suggested the problem lay in the Swarthmore honor system! He replied if I thought that sort of thing is bad, just imagine what it is like in a psychology department. This stimulated him to go on to make an illuminating mini-speech about the problems of being a graduate student. He said that it was a period in one's life where one would normally expect to be totally independent, yet not only is that not the case but one's professor holds the power of life and death. The result is that the student, totally unconscious of what is happening, will begin to imitate the way of speaking, the way of dressing, and in many other subtle ways the attributes of his all-powerful sponsor. It is an automatic defense mechanism that overtakes the student without his realizing it.

It did not occur to me at the time but after some years I realize that this is exactly what had happened to me. In those days I could not have imagined it to be true for Cap was such a thoughtful and kind person, but now I know that he entered my skin in ways I could never have imagined possible.

By the spring of 1947 all universities had too many students and too few faculty because of the GI Bill that paid for the vast number of veterans' educations. It could not have been a better time to matriculate. I could have stayed at Harvard another year because I had my fellowship (a Junior Fellowship that was well paid and prestigious, giving my career an early boost), but I decided I wanted to get on with my life. The war had held me back long enough and the family was growing; Ruth was pregnant with our second child. There were three places that invited me for a job interview: Amherst, Johns Hopkins, and Princeton. I decided Amherst was where I wanted to go, but it was the only one of the three that did not make me an offer. Princeton was my second choice, and it turned out to be the perfect place for me.

Fortunately I have kept the letters offering me the job at Princeton. My acceptance is particularly revealing in that it shows how extraordinarily modest were the requirements for doing biological research compared to what exists today. Here is my letter to the chairman:

> Dear Dr. Butler,
> I should like to formally accept the position you have offered me (in your letter of December 18) as a Research Associate with the rank of Assistant Professor at a salary of $4000. I consider this offer an honor and it is with the greatest pleasure that I accept.
>
> In our telephone conversation on December 27 the major questions on my mind were answered. There are a few minor and more material ones that occur to me now . . . [After discussing when I would start, I continue:]
>
> Another question I wanted to bring up was basic research equipment. You undoubtedly have all I need already, but on the chance that you might not, it seemed wise to list the items of equipment, besides glassware and chemicals, that I would use constantly: compound microscope, camera lucida, dissecting scope, autoclave, dry sterilizing oven, small incubator, small centrifuge, tripple [sic! I'm surprised I didn't put in three *p*s] beam balance. If those basic items were available then I could dive right into the research the moment I got there which would be most desirable . . .

Today new faculty members could not possibly begin without a large "starter grant" to buy needed equipment, involving many thousands of dollars. They would still need the microscopes but also a computer and many expensive devices for the molecular aspects of his or her (another innovation!) work, which is indispensable nowadays if one is doing experimental developmental biology. It is possible that a young biologist now has little idea what a camera lucida is. It is a device that fits over the eyepiece of a microscope consisting of a prism and a mirror sticking out on the side. If one looks through the microscope one can simultaneously see the object under the scope and one's pencil on a piece of paper to the side. In this way one can exactly trace the object, giving its precise dimensions and proportions. Today one uses all sorts of clever cameras and even time-lapse video attached to the microscope, which makes such operations infinitely easier. Learning how to use a camera lucida took time and patience and (for very good reasons) has become a lost art in biology.

Not long after I started at Princeton, the National Science Foundation was established and it became possible to apply for grants. This was particularly important if one wanted to hire a research assistant. I applied for what today would be considered a ridiculously small grant, and in the letter that told me I was successful the NSF asked for an annual report in the form of a letter. After the first year I wrote that things had not worked out very well—I had tried this, that, and the other, and nothing had really worked. (Can you imagine writing such a letter today?) They wrote back saying, "Don't worry about it—that is the way research goes sometimes. Maybe next year you will have better luck." (Can you imagine the NSF writing a letter like that today?) So with all the wonders and marvels of our progress in laboratory biology during the last fifty years, there is a price we have had to pay. But for many kinds of experiments it could not be any other way. It may be fun to reminisce about the good old days, but it is far more rewarding to admire the truly extraordinary changes of the intervening years.

A short time ago I was asked what the Princeton biology department was like in my early days there compared to today. In the first place it was small; there were ten faculty members—I was the eleventh. Today we number in the fifties and are split into two departments. Then it was like joining a club—now it is a professional institution. There were very few graduate students, and the undergraduates took to their studies in a much more relaxed fashion than they do today. Reminders of the undergraduate days of F. Scott Fitzgerald were still visible, although the more mature returning servicemen were beginning to change that. And there were fewer undergraduates: everything is larger today.

One bit of early excitement was the fact that soon after I began I had a senior and two graduate students doing research with me. The graduate students were a new breed: they too had been in the service and were back at the university, just the way I had been only a short time before. So we were close in age and happily got along well. I had told them one day that I had gone directly from buck private to Officer Candidate School, and I had always admired master sergeants as the ones with the supreme power. I used an old fatigue jacket in the lab when doing dirty jobs, and one morning, to my delight, I found that master sergeant stripes had been secretly sown on during the night; what more could I achieve!

One of those students, and the very first to get his doctorate with me, was David Stadler, who later went on to a distinguished career at the University of Washington in Seattle. One day David's father came to visit his son and I was asked to join them for lunch. This put me in a bit of a state because L. J. Stadler was one of the really grand men of genetics at that time—what would he think of the very young man teaching his son? I should have known better because David's father was indeed the splendid person I had been told about. He had the gift of making one feel that the privilege was his, although we both knew perfectly well it was the reverse. Erwin Shroedinger's *What Is Life?* had just been published, and we had a stimulating discussion during lunch about what it all meant and what it held in store for us. This was a revolutionary book by a famous physicist that in many ways heralded in the new age of molecular genetics. In a slim volume Schroedinger managed to pose the key question: what were the properties of genes that give them such remarkable capabilities? As a theoretical physicist he only asked the very general question but his book was a great catalyst that contributed to the explosion that soon followed, first with an understanding of the genetics of viruses and then the structure and properties of DNA.

In the beginning I had students working on the development of numerous different lower organisms, algae and fungi particularly, but as my own work on slime molds progressed I found that more and more of my students wanted to work on them too so that my laboratory slowly emerged into a "slime mold lab." There are a number of advances we made during that period that were rewarding but let me describe just a few to give some idea of why I became so engrossed.

Each year I would have some seniors doing research with me for their senior thesis, along with graduate students and occasional postdoctoral fellows. In the early 1950s a senior was going to look into the orientation toward light of slugs in the migrating stage to see what colors were particularly effective in attracting them. The idea was that in this way one could get some idea of what kind of pigment within the migrating slug was absorbing the light and causing the orienting-to-light response. (Years later this goal was successfully carried out by others and is now well established.) My student put the culture dish containing the slime mold slugs in small wooden boxes in which a window was cut at one end. In these windows he pasted with electrical tape different color filters and placed the boxes some distance from a fluorescent light in an incubator. The next morning it was easy to see the direction of movement of the slugs because they leave visible tracks behind. The only problem was that he found that no matter what color he used, the slugs crawled toward the light, which to me seemed impossibly puzzling. He then tried something that would never have occurred to me: he placed the chalk boxes backward so that their light-proof end was toward the light bulb, and the next morning he came rushing in to my office with a "eureka" look all over his face. The slugs still moved toward the light even though they could not see it; therefore, he explained to me, they must be responding to temperature gradients. He quickly tested this in a few days and was able to show that slime mold slugs are indeed incredibly sensitive to heat gradients, so much so that we were able to estimate that a small slug would turn in the warmer direction if the difference in temperature between its two sides was as little as 0.0005°C. When I would describe these experiments in seminars no one believed them, and that taught me a lesson: if one's results are met with disbelief there is a good chance that one is on to something really new. I am happy to report that these results were not only confirmed in the 1970s in another laboratory (using wonderfully sophisticated apparatus) but they went on to show some other genuinely interesting aspects of thermotaxis.

Another step forward was the discovery that individual amoebae do not always maintain a fixed position in the migrating slug but they can and do move about. The first inkling of this came from making grafts with slugs that were stained with a vital dye. If one grafted a segment from the anterior end of a red slug into the posterior end of a colorless slug, the red cells would move through the colorless cells and end up in front. The fast cells percolated through the slower ones. In itself this was just a curiosity but then I found that in normal development in any one species there was a "sorting out" within the cell mass, and the faster cells went to the front end while the slower ones lagged behind. This was quite contrary to expectations; it had always been assumed that the first cells to enter an aggregate ended up at the anterior tip, and the last cells were at the tail end. This sorting out of cells has not only been confirmed by others but explained by them as well: the cells with the largest food reserves are most likely to end up as spores while the leaner amoebae are likely to become stalk cells, which makes good sense from the point of view of an evolutionary strategy, for the cells rich with food were the safest candidates to start the next generation. Much later my last graduate student showed (before she came to Princeton) that cells that were starved just before they divided (and hence replete with food energy) tended to become spores, while amoebae that had just divided when the food had gone tended to become stalk cells.

Shortly after I arrived in Princeton we were invited to a cocktail party and I met Paul Oppenheim, a charming German philosopher of science who told me he was very interested in biology and asked me what I worked on. He pressed me even though I told him I worked on very curious amoebae, and when I told him they were slime molds his face lit up and he began to tell me all about their peculiarities. I was stunned and asked him how he knew this because at that time even most biologists did not know of their existence. He said that some years ago in Germany in the 1930s he had heard a lecture by Arthur Arndt, who showed a wonderful time-lapse film of their development (and it is indeed a fascinating film; I still have a copy). He went on to say that the lecture was quite extraordinary because Arndt said that the life cycle of these slime molds was so amazing that any materialistic explanation was out of the question; it could only be explained by some mystical vital force. There is no hint of this in Arndt's 1937 paper on the organism, but he apparently felt no constraints in a lecture.

After that Paul Oppenheim and I became friends, he always asking me about my experiments and at the same time trying hard to make a philosopher, a logical positivist, out of me. He was a close friend of Albert Einstein and that was the main reason he eventually chose Princeton after he had to leave Germany. One day he called me up and said that Professor Einstein would like to see my film (my old senior thesis film) that he had told Einstein about, and would I come to Einstein's house to show it. Of course I said yes, and arrived one afternoon with the film and a projector. We had trouble finding a suitable screen but finally turned a wall map of the United States around. We were joined by Miss Dukas, Einstein's formidable secretary. After the viewing Einstein asked me if I would come in to his study to discuss what we had seen. We talked for some time and I have always cursed myself for not having written down what I remembered of the conversation right away in a Boswellian fashion. I do remember that his questions went to the core of the problem of development—something I have been pursuing all my life.

I also remember how kind and gracious he was to me, a very young and callow aspiring scientist, and he had no trouble understanding my English. Occasionally he would stop and think and one of the others would assume he had not understood, so would repeat what I said in German. Each time this happened he got quite testy and said of course he understood. After we stood up to go I told him I knew the philosopher Alfred North Whitehead and had once asked him if he had ever met Einstein. Whitehead replied that indeed he had—under the most embarrassing circumstances. Lord Haldane, a very forceful man, had invited them both to dinner, after which he escorted them to his study and left them alone, saying they must have so much to say to each other. He told me, "Both Professor Einstein and I are very shy men, and we had an excruciating time—neither of could think of what to say." I asked Professor Einstein if his memory of the event was the same. He gave a warm smile and said it certainly was—it was very painful indeed. "You see," he said, "I was never able to understand anything Whitehead had ever written, so what could I say?"

As we were about to go out the front door someone rang the bell. Miss Dukas pushed Professor Einstein behind the door and opened it a bit to see who it was. She closed the door and reported that some professor of physics wanted to talk to him. He said, "That old fool—tell him I'm busy."

During those early years at Princeton, and especially that summer in Dr. Conklin's laboratory at Woods Hole, I began writing a book on developmental biology. I wanted to show that the methods and the ideas of the old embryology could be extended to all organisms, and that bacteria, algae, fungi, amoebae and slime molds, and protozoa did as much developing as animal embryos used in conventional embryology. My book progressed and after many revisions incorporating the helpful comments of friends, I finally submitted it to Princeton University Press. I found the whole process of getting the final version ready and sending it to the Press tremendously exciting. I always had a great desire to write a book, for reasons I never fully understood. It was sort of a primitive feeling that a book would be the ultimate in accomplishment. The readers' comments for the Press sent me into an absolute dither: each compliment became etched in my mind, and each criticism was a deep wound. After I simmered down I took all the criticisms seriously, made many changes, and I knew the manuscript was better for it.

Each subsequent step was a magic event. First the editorial board accepted the revised manuscript. The next, even bigger, thrill was the galley proofs. How glorious it was to see my words set in type. I had seen my prose before in journal papers but somehow this was different. I went over the galleys for mistakes with loving care, knowing there was little chance that I would see them through my rose-colored glasses. Then the page proofs, with a title page—each step was from one cloud to another. The dust jacket came before the book; I had never seen anything so beautiful, with its title *Morphogenesis* and my name, all in bold print. When the first finished copy arrived it was difficult to contain myself.

Over the years I have often thought about my reaction to the birth of my first book, a reaction that was in many ways remarkably childish. Yet instead of finding it embarrassing, I still wish that I could respond to things that way again. Experience, despite its many advantages, unfortunately makes one jaded. That primordial thrill can never be fully recaptured. Some years later it happened to me all over again with my first salmon: I was fishing for trout and a salmon took my fly and was finally grassed many minutes later—oh, those palpitations.

After the book came out, I had the feeling that no one paid any attention to it, and that the book really did not come into its own until ten years after it was published. I have checked this recently in my files, and there is no truth to it; the reviews were quite favorable—it just took me ten years to relax enough to understand what had been the outside reaction. Perhaps the main reason for that was a partially suppressed conviction on my part that I could not write. At school I had to take remedial writing during the summer; Cap Weston deplored my "pure Bonner" prose; the few things I wrote and showed friends were met with comments that clearly implied disapproval. Yet for some reason I was determined to write, and the first encouragement came as a freshman at college in a compulsory English course I had to take because they said I needed remedial work. My teacher told me sternly that what I wrote about was the pits—all this stuff about the beauty of the woods with the birds singing—why did I not loosen up? At the time I was reading James Joyce's *Ulysses* so my next effort was the first chapter of a book on student life written in the manner of Joyce. My grades jumped from Ds to As and my instructor actually smiled at me. I went on taking writing courses, one with Robert Hillier that was particularly enjoyable, and even had a poem published in the student literary magazine, and two articles on doubtful science in the Sunday edition of the *Boston Herald*. But I still had inner doubts, so when I read in the reviews that my writing style met with approval I was stunned.

After the Second World War my father had quite a splendid position in the United States embassy in Rome. Both Ma and Pa loved the high life of the "Corpo Diplomatico." They lived in a sumptuous and elegant apartment in the Palazzo Torlonia near the Spanish Steps and, of course, knew everyone. Ruth and I visited in the summer of 1949, and I have many memories of a different world. One image flashes through my mind that somehow epitomizes their life and that summer. In the guest bathroom the tub was out in the middle of the floor. As one reclined in the tub one looked up at a rather splendid eighteenth-century fresco with cherubs and clouds on the high ceiling.

After he finished his tour of duty there, the State Department asked my father to continue but he decided that he wanted to resume his interest in writing fiction. Until then he had written mainly short stories; now he wanted to try a novel. He and Ma returned to America and eventually bought a house in a small town outside of Charleston, South Carolina, and he began writing in earnest.

His first novel was called *S.P.Q.R.* and it had the high life of Rome as its setting. It was an immediate success and quickly appeared on the *New York Times* best-seller list. We were overjoyed for him: here was yet another new career and again a successful one. He was almost sixty years old when it came out; it was the first of a series of novels, each one of which was a success.

Partly no doubt because of mutual exhilaration at having published books the same year, and partly from paternal pride, he and I used to have these wonderful between-authors conversations in which inevitably came the question "How many copies of your book have been sold, son," to which I would reply that it was doing very well: it had sold 378 or something of the sort. Then when I would ask him the same question, we were suddenly dealing in the thousands. Fortunately it was a bone of amusement rather than contention—we were both riding so high we were incapable of feeling badly about anything.

There was another related incident that occurred later. This was told to me by my mother who thought it was a great joke. It has to do with *Who's Who*. First I should explain that *Who's Who* automatically includes all permanent faculty members from research universities; it was a matter of no great moment. Pa was asked to send in material, which pleased the new author very much, and he bought the volume with his entry in it. To his amazement (and his delight) he discovered his son was there too.

This story spans three generations. At roughly that time our son Jonathan was in nursery school, and one day the teacher asked the children, around in a circle, what their fathers did. (Note that in those days there was no question of what the mothers did!) Jonathan reported that Julie's father was a policeman, Jimmy's was the fire chief, Sally's a doctor, and so on. He was telling this to Ruth, so she asked him, "What did you say your father did?" He replied, "I had to tell them he didn't do anything."

My work in the biology department at the university was a joy from the beginning. I enjoyed teaching and I was able to get my research done without difficulty. I was told that at the end of five years I could have a semester off for a leave, and as the time approached I decided to try and get a fellowship to go abroad. I applied to both the Rockefeller and the Guggenheim Foundations with the understanding that I would take the first one that came through. This happened to be the Rockefeller, so I quickly wrote to let the Guggenheim people know. What I did not realize was that the Rockefeller stipend was alarmingly small because the Foundation did not approve of families going as well. The money provided was enough to support only me, but I was taking Ruth and the children.

We decided to go to Paris where I wanted to work in the laboratory of Professor E. Fauré-Fremiet. It was a fortunate decision because Monsieur Fauré, as he was called by everyone in his lab, was a winner on all fronts. I had admired him for his published work on the development of ciliate protozoa and asked him if I could spend seven months in his laboratory. I soon found out that over and above his science, he was an absolutely splendid person, much beloved by everyone working for him. He was a short man, straight-backed and with a brisk manner who displayed a fine combination of charm and wit. At the time I was there he was seventy but he seemed much younger simply because he had such spirit. We became good friends: it was rather a father-son relationship, although I could have been his grandson. Quite often he would come into my room and say let's take a little walk, which meant taking a stroll around the circular hallway. He had many photographs of friends and fellow scientists lining the wall and we had a mutual acquaintance, Leigh Hoadley, my professor of embryology who had worked with Fauré some years before and whose youthful photograph was among them. He would stop and point to it and say, with a great smile, "Ah, voilà le Hoadley. Qu'il est beau!" and then we would proceed on our walk and talk as though there had been no interruption.

Fauré had no children of his own. Madame Fauré and her sister kept house for him. They were the daughters of his predecessor at the Collège de France, and they always seemed to me like two kindly but rather sharp old spinsters. He had a keen eye out for pretty women but in a nonthreatening way. A number of the laboratory assistants were knockouts, and he referred to them as "mes petites." They adored him, as did everyone else. One day I knocked on his door and he immediately said, "Entrez." I opened the door and there he was beaming at me with his arm around and snuggled up to one of his petites, who beamed at me too. There was absolutely no embarrassment. It could not have happened in an American or a British laboratory. There was certainly nothing macho about him; furthermore, he was a very modest man. He never mentioned the fact the he was the composer Fauré's son, the grandson of the sculptor Fremiet, and the nephew of Sully-Prudhomme. Once when I remarked on a elegant drawing of a protozoan he had made, he explained that he had wanted to be an artist as well as a biologist, but his grandfather told him very firmly he must make up his mind—he could not be both.

I had wanted to work with him because he too was interested in the development of lower forms: in his case it was with ciliate protozoa (such as *Paramecium*) with which he had made some important and interesting discoveries on how they developed, and especially how their complicated cortex with its cilia and reticulated structures reorganized after cell division. It seemed to me at the time that it would be valuable for me to work for a while on an organism that was quite different from my own—otherwise I might get a distorted, slime mold view of the world. I realize now that my decision also probably reflected my deep-seated fascination with life cycles. Ciliates were often very large, yet they were single cells—how could they be both? The answer was a very interesting one: they do all sorts of strange things both with their nuclei and the outside covering of their cells to make it happen. Their life cycle was as odd as that of slime molds, yet also totally different from that of most animals and plants. It turned out to be a good decision from many points of view; in particular it greatly expanded my understanding of how and why organisms develop.

As I look back over my years as a biologist I can see that, beginning with my exposure to lower forms in Cap Weston's course in my freshman year, in the algae course in Woods Hole, and into my early research on slime molds, I became increasingly fixated and fascinated by the fact that not only do all organisms have life cycles but the variety of those cycles is immense. There was no conscious plan on my part to lay out a program of study along these lines; it just developed within me as though I had been programmed by unknown forces. There is no doubt that all biologists gain much by becoming intimately knowledgeable about one organism, and slime molds certainly have played this role in my life. But more than that, they have forced me, seemingly without premeditation on my part, to look at other organisms in a parallel way. My first step in this direction was this foray into ciliates.

Working in Fauré's laboratory provided me with an unrelated and unexpected new lesson. I had to make some microscope slides of my ciliates and Monsieur Fauré said he would get one of his "petites" to teach me. My first difficulty was that I had written the whole procedure down and when I followed it slavishly the results were not very good. My coach told me that was not quite the way to go about it, and in half the time she would get a perfect result. This went back and forth for some time, and I finally realized she was approaching the technique as an art, while I thought of it as an exact science. Furthermore, hers was the art of French cooking, while I had been brought up in the American tradition, directly inherited from the nineteenth-century German laboratory—the tradition of precise measurements. I was also taught how to keep my ciliate cultures: take about so much wheat grains in the palm of one's hand and add about so much pond water and boil for about ten minutes. It worked perfectly; I got so that I was sure if I had carefully measured them out, everything would die. Laboratory science was an art.

I can think of another example that told the same story. I visited a laboratory where there was a charming old man who was a direct descendent from Louis Pasteur's laboratory. He wore a black skullcap, the kind Pasteur wore in some of the old pictures. I told a friend of my visit and he said he had made a similar one, and he asked the man how he grew some bacterium that was hard to culture. The answer was that he grew it on agar that was "légèrement maltosé," which roughly translates to "lightly malted." Clearly in France first-rate science can go hand in hand with marvelous cuisine.

Finding a place to live in Paris, we had been told, was impossible. Long before we left I had sent out innumerable letters, with no result. I mentioned our predicament at a cocktail party in Princeton, and someone said to me they had American friends who were just vacating an apartment in Paris and I should get in touch with them. I did so immediately and in no time at all we had the perfect place on the Rue de Bourgogne, half way between the Rodin museum and the Chambre des Députés, which is in a beautiful part of the city. Not only that, but it came with Marguerite, who turned out to be a splendid person and a fantastic cook. We never had nor will we ever again eat so splendidly as we did that time in Paris. The irony was that because of the Rockefeller no-family policy we had never been quite so short of money. We lived like a royal family, yet we had to have a daily evening conference, poring over our small collection of francs, to decide what we could and could not afford to do the next day. I got paid every two weeks: I would have to go across town to the Paris Rockefeller office, where the cashier would hand me a wad of bills held together with a straight pin. In a sort of unctuous manner he would ask me how we were managing. I soon realized that since I had broken the no-family rule, he was always hoping I would say we were starving. To play the game, each time I would assure him we were managing famously.

By that time we had our third child, Jeremy, a toddler still in diapers. Marguerite referred to him as "mon petit crapaud" as he crawled about the floor. The diapers presented one of our first domestic crises. The first week I took them down to the laundromat and the only way they could estimate the price was by weighing them. I pointed out that dirty diapers were bound to be wet and heavy; surely I could count them out. No luck—the first load cost a fortune. At that rate we would indeed soon be starving. To solve the problem, before going to the laundromat I would put the rinsed-out diapers on the roof, outside the dormer window— they looked rather like drying codfish in Newfoundland.

The older children went to the local school, the école maternelle, where they learned French remarkably quickly. Jonathan, then five, was picked on quite a bit because he did not cotton to the fact that the teacher believed the first child who snitched. Every day there were wild tales of class. "Marie Françoise stole my gomme [eraser] so I socked her, and she told the teacher, so I got put in the corner." Rebecca, then eleven, discovered the English lending library and read nonstop, loving every moment. We hired a woman to take the children out in the afternoons. She was an attractive woman in her forties who must have been ravishing in her youth. She was quite brainless, but the children liked her and referred to her as "Madame Ooh-la-la." The only person who disapproved was Marguerite: the baby-sitter had said that when younger she was an "artiste" and Marguerite would shake her head and say in dark tones, "You know what that means."

We took many excursions and were very busy sightseers, always as the whole family. Sainte Chapelle is etched in my mind: I was carrying Jeremy and he suddenly discovered if he let out a baby whoop, it echoed. I had never heard such a racket, much to our embarrassment and the amusement of some of the other visitors. One beautiful day we were reveling in the park of Versailles. For some time I had been furious with the stroller we had brought from Princeton; it was piece of junk that kept breaking down. In the garden near us I saw a man with the ideal stroller, with large sturdy wheels. I rushed over to him and asked him in my best French where he had bought it. He smiled and replied in his best American that it came from Marshall Field's in Chicago!

My youngest brother, Tony, and his wife, Eve, whom we met for the first time, were living in Paris. They were both beginning expatriates: Eve had been studying at the Sorbonne when they met, and Tony was doing music composition under Nadia Boulanger. From the very beginning it was wonderful to have them there for we all got along famously. They were great experts on what ancient monuments and museums to visit and, even better, by careful research they knew all the best cheap restaurants and Tony furthermore would tell us exactly what to order. They greeted us when we first arrived in January, frozen and starved, and took us to a tiny restaurant and told us to order tournedos in Armagnac sauce, whose taste I can remember to this day. The restaurant was kept warm by a glowing potbellied stove in the middle of the small room. We took a table near the stove and soon we were all warm and full, ready to attack the world.

Tony also continued to work on my musical education, which he had started a few years back. A high point was when we all went to hear Stravinsky's *The Rake's Progress*, an opera I have loved from that moment. Nadia Boulangier and Stravinsky were there—it was a festive occasion. We did not have enough money to go to many concerts, even though things in Paris were quite cheap at that time, but for a birthday present Ruth gave me a ticket to *The Magic Flute*. The title sounded much better in French: all the posters said *La Flûte Enchantée*. It is an opera that has a strong emotional effect on me. Nothing could be that beautiful.

Ma and Pa came through for a brief visit, which produced a great bustle to get the right wine, and the dinner required great preparation; Marguerite outdid herself. Afterwards she, as she often did, felt compelled to comment on our guests. If they were Americans, not much was said, but my parents were different. She was particularly taken by my mother, and the next day she pronounced the ultimate compliment. She told Ruth that "Madame est une grande dame." One evening we invited the Fauré-Fremiets and Madame's sister. When we told Marguerite, she was in a state—the guests were French! She would not let Ruth do any of the shopping; she decided we were to have duck and she went to the market and so far as I could determine spent some time squeezing all the ducks there to get just the right one for the particular dish she had in mind. The dinner was superb but had to be done in the French fashion with a succession of separate courses. We did not have enough plates and silverware so we had to hire the concierge to wash dishes nonstop between courses. After dinner Madame Fauré and her sister went into the kitchen to compliment Marguerite and give her a tip. The next day when we were thanking Marguerite for the big success, she made it very clear, in a nice way, that she very much approved of our French friends. There was hope for us yet.

To go to the laboratory from our apartment was about a half-hour walk. Even if I took the bus I would get off at the Luxembourg Gardens and walk through them. To this day I can still see those beautiful spring mornings when the clipped chestnuts were just turning green and the thin sun barely warmed the ground. Some mornings the gardeners would be rolling out the potted trees from their winter quarters in the Orangerie, a process that no doubt had been carried out for the last two hundred years. In the chairs by the fountain there would be a few students getting a short drink of sun before a morning of dusty classrooms. In one corner there was an eighteenth-century statue of a nude woman—her figure was so exactly Ruth's I felt somehow I should rush over and cover it. The last lap to the laboratory was down the somewhat severe Rue des Écoles and then into the grilled courtyard of the Collège de France. Every day I was aware of the plaque on the corner of the building that said this was the laboratory of Claude Bernard, and, if I missed that, his statue was out front facing the lodging house where he spent his later years. I never heard an outsider give a lecture in physiology at the Collège de France without saying what a privilege it was to give a lecture in the institution of Bernard. His presence was almost felt. There were also the names of other famous scholars inscribed on the walls but more poignant were the names of the professors and the assistants who were killed during the Second World War. I no longer remember the names but some of them were women, and under the name of one was "Tortured and killed at Auschwitz."

I met a number of distinguished scientists, many of them introduced to me by Fauré. He particularly wanted me to meet his old student, Boris Ephrussi. I had heard that Ephrussi was rather a difficult and domineering person, but I thought I might be safe if I went with his old teacher. I soon found out how wrong I was. Ephrussi's English was better than mine, but I suppose in deference to Monsieur Fauré, the attack took place in French. I had just published *Morphogenesis*. His first question after a polite handshake was what did I think might be the answer to the problem of morphogenesis! I had spent a whole book trying to answer the question in English, and the idea of giving him a sensible answer in a paragraph in French was not likely to happen, as we both knew. He flattened me in the first round! Later he came to Princeton and we had some very reasonable, if slightly edgy, conversations. The only person who did the identical thing to me quite a few years later was Max Delbrück, an equally distinguished biologist who was one of the important contributors to the beginning of molecular genetics. I was able to handle that a bit better—I had been through it all before. How to rattle someone in the beginning of a conversation so that the top dog can be quickly established.

I was also introduced to another remarkable group from the Pasteur Institute. An informal seminar, including Ephrussi and a number of others, met periodically in the late afternoon to discuss modern problems in developmental biology. It included Jacques Monod and André Lwoff who, along with François Jacob, did so much to further our understanding of the way gene activity is controlled in bacteria and viruses. Particularly important was their discovery of how the activity of a gene is controlled in the bacterium *E. coli*. They were major players in the beginnings of molecular biology, for which all three shared the Nobel Prize. As I look back on those seminars in 1953, I realize that already these first-rate people were plotting the molecular revolution. I was intrigued by what they were doing and were reporting at the seminars, but I felt at the time that there was a great gap between their studies on gene action and my more global interests in how an animal or a plant develops. Over the years that gap is slowly narrowing, but at that time their vision into the future was extraordinarily prescient.

In that group the person with whom I felt the greatest affinity was André Lwoff, who was the head of that particular Pasteur Institute laboratory. This was partly because he had written an excellent book jointly with his professor, which I had read previously, on the development of ciliates. It was totally biological in its content and full of interesting ideas on how the structure of their elaborate cortex was inherited from one generation to the next. I decided to call on him at the Pasteur Institute but unfortunately he was not in, so I left a note saying I would very much like to meet him and left a copy of my book. Very soon afterwards he telephoned me to say that he had not known that I had produced such a major work, and he would be delighted to see me. He stopped almost in mid-sentence, however, and asked, "Where did you learn to speak French? I suddenly realize I am talking to an American who can speak French!" I told him about my Swiss background but warned him that soon he would see cracks in my syntax that I would try hard to conceal by speaking in short sentences.

That deficiency became all too evident when I gave a lecture in French. I cannot read a lecture to an audience because I have never been any good at reading out loud, even my own sentences. I lectured twice and both times it was an effort—I felt as though I had been engaged in a one-man wrestling match. After a very formal lecture at the Collège de France my friends in the laboratory came rushing up to tell me all the mistakes I had made in my French. They told me I should not despair—some of them were very funny!

The group at the Pasteur Institute and Ephrussi and Fauré seemed so different from many of the other scientists that I encountered in France that year. They were, along with some of the older, more established biologists I met through Monsieur Fauré, international scientists who kept in active touch with what was going on in other countries; they were very much in the forefront. I spent even more time with the younger people in the laboratory who were closer to my age, and some of them became good friends. It was surprising to me at the time how insular they were. They kept in close touch with French science and the French journals but they paid much less heed to what was going on elsewhere. No doubt part of it was a language barrier and perhaps part of it was indoctrination at school. Surely that is no longer the case, what with English becoming the universal language of science; in France today there is an increased awareness that science has little interest in borders.

As I look back on those months in Paris, I have nothing but pleasant memories, although I know all was not smooth. One big political event that gripped us all to the bone was the execution of the Rosenbergs for espionage back home. I found out later that most people in America felt they were guilty and should be executed, but in Europe in general, and in France in particular, there was a very strong feeling of sufficient doubt of their guilt so that their lives should be spared. It was the first time I realized that a country could be whipped up into a passion on a particular point of view, one quite different from the feelings of those in another country, all based on the same facts. We were genuinely upset as the date of execution approached, as were all our French and American friends. All over Paris there were huge posters with a cartoon of President Eisenhower grinning, and each one of his teeth was an electric chair.

There was a very nice young man in the laboratory who decided to make up a petition that we would all sign, and he would take it to the American Embassy. He did the rounds and was going to take the petition in after work. Suddenly, at the very end of the afternoon, Monsieur Fauré came into my room and explained that he was collecting signatures all over again, and he had persuaded the nice young man (whom he described as a "chic type," which is roughly equivalent to "nice guy") that this was in the best interest of the Rosenbergs. Fauré explained to me that since our friend was a communist, as of course we all knew, the previous petition would do more harm than good. Naturally I signed again and Monsieur Fauré rushed out to get a taxi to take it to the embassy. We did not save the Rosenbergs, but our respect for Monsieur Fauré went up yet another notch.

When I returned to Princeton I began to ask myself questions, such as why do we have development at all; why go to all that bother of starting as a single cell in the form of a fertilized egg and each generation constructing a large complex adult? Would it not be easier for an elephant to simply split in two and the front half regenerate a new posterior and vice-versa? Some worms can do this, but they also have a complete life cycle that starts with a fertilized egg. It slowly dawned on me that this was something that had arisen, and was maintained, by Darwinian natural selection. To compete successfully there must be inherited variations, and sexual reproduction is the effective way of handling and disseminating that variation. In a multicellular organism all the genes of all the cells are the same; each cell has the complete complement of the organism's genes. In every generation there is a mixture of the genes of the two parents, and this can only be achieved by the fusion of one cell from the father (sperm) and one from the mother (egg) to form the new offspring. In other words, in each generation there must be a single cell stage for sexual reproduction to take place. It is interesting that even asexual organisms, which include many algae and fungi and some invertebrates, also commonly have a single cell stage in their life history. The advantage there seems to be largely for dispersal; small single-cell asexual spores can effectively be spread by the wind and other agents.

We also know that there has been, over the course of evolution, a selection for size increase—under many ecological circumstances a larger organism will have an advantage in the competition for resources over a smaller one. So natural selection is simultaneously favoring a very small stage for managing heredity or dispersal and a very large stage for effectively competing for energy in the form of food. This led me to an important conclusion about the life cycle: that development was the inevitable result of sex and size. The life cycle was framed by these twin pressures of natural selection.

Obviously these ideas were a direct outcome of my fixation with life cycles. Slime molds, ciliates, algae, insects, human beings—we are all life cycles. It is not just the adult that evolves through natural selection but the entire cycle. We are so naturally inclined to think of ourselves as adults that we neglect the fact that we began our development as a single fertilized egg. We automatically think of an individual as an entity that exists in a moment in time—a snapshot—when it exists in a particular shape; in human beings that moment could be when we see a fetus or when we see an old woman. However, that is a trick our brain plays on us—from the point of view of evolution an individual is a life cycle. Natural selection culls or encourages every stage of our life span; in fact the only way to change the adult is to make a change in some earlier period of the development, of the life cycle.

Just at the time in 1956 that some of these thoughts were taking shape in my mind, I received an invitation from G. P. Wells to give a course of three lectures at University College in London. I was excited and pleased, and immediately began to put some of those grand thoughts together in the form of lectures. I had written up the main ideas, and now I rewrote them completely.

The whole event was a big moment for me. I felt as though I was finally coming into full bloom but at the same time I was terrified. This was not helped by the fact that as I took the train from Princeton to New York to go to the airport, the train came to what seemed like a permanent stop when the drawbridge over the Hackensack River was stuck and the tracks could not be lined up properly. After an agonizing delay we finally got through, and I just caught my flight.

University College put me up at the Ciba Foundation on Portland Place, which was supported by the Swiss pharmaceutical company. There were a number of other scientists from all over as guests, and at one memorable breakfast an Englishman appeared, scowling at everyone. He surveyed all the beautiful Swiss jams and said in a furious tone, "No marmalade—hardly an English breakfast!"

I had splendid reunions with Clare and Justin Lowinsky, who had spent the war with us as refugee children and were now grown up, and the rest of their family, and Brian Shaffer from Cambridge, a good friend whose work on slime molds I particularly admired—among other things he was the first to isolate the chemical that attracted the amoebae. I could hardly believe it: the very first evening the Ciba Foundation put on a dinner for me, after which I gave a lecture on my recent work on slime molds that was chaired by Peter Medawar, and there were other distinguished biologists there whom I knew only by name. Nothing like this had ever happened to me before—I felt like a debutante. And the main show had not yet started.

The day before the first lecture I was so nervous that I decided it might be wise to go to see a play or a film for therapeutic distraction. I found that Agatha Christie's *The Mousetrap* had a matinee—the perfect medicine. It had already been running for years. I was the only man in the audience, and at the intermission tea was served. I was totally distracted.

The first lecture was even more terrifying than I thought possible. I was not allowed to just enter the lecture room but was marched there by the most magnificent beadle, all dressed in a light blue uniform, rather like the doorman at a very fancy hotel. He carried a huge mace as we proceeded to the lecture hall—I was convinced I was marching in my own funeral procession. We went to the podium where I sat next to Gyp Wells who introduced me. The large lecture hall was full, and in the front rows were people well known to me and whom I had just met. In the front row I could see a star among neurobiologists, J. Z. Young, and not far from him the famous geneticist, J. B. S. Haldane, who positively glowered at me. How I got through the lecture I will never know. Just afterwards I rushed to the Gents and as I was washing my hands, Haldane, who was next to me, said in his booming voice, "Bonner, we don't make jokes in our lectures in this country." This did nothing to calm me down but I did manage to say, "Those weren't jokes; I was just nervous."

I had no idea how the first lecture was received, and walking down the street the next day I ran into J. Z. Young, who greeted me with his charming smile as he inquired, "Well, John, what did you think of your lecture?" I always thought that was a splendid ploy. Later one evening we went to a series of pubs drinking beer, and while he never said so, he made me feel that my lectures were not a disaster.

The same was true for Haldane, and he insisted I come to dinner with him and his wife, Helen Spurway. They both loved to shock and they loved to argue. Helen had just been arrested for a misdemeanor involving someone else's dog. I forget the details but it was in all the newspapers, and she was reveling in it and the principle she upheld, whatever that might have been. We went to a small restaurant in Soho and soon were embroiled in some very spirited arguments. One was about some biological aspect of sex. Mostly they argued with one another, and both of them had very penetrating voices as they become more intense, with the result that all the people at the neighboring tables were staring at us. Even though they both appeared to ignore the stir they were causing, I could not help feeling they not only were aware of it but enjoyed seeing the shock waves travel across the room.

By the time we left the restaurant we were on to the recent work in animal behavior and its evolutionary implications. It was a subject of concern to all three of us, and indeed it was the central theme of one of my lectures, where I drew parallels between behavior and development. We all had more to say so they decided they would walk me to Portland Place, but we still had not finished so I walked them back toward University College. The whole process repeated itself again before we were ready for bed. On one of the laps we passed the BBC building and on the ground floor a low window was open at the top, and one could hear a radio blaring away. Haldane was talking, and suddenly he veered across the broad sidewalk, stood on his toes, shoved his enormous head into the open window, and yelled with tremendous force, "shut up!" He went directly on to his next sentence as he cruised back alongside us without skipping a beat. I got the impression he did not like the BBC! And I always wondered what might have been the sensations of the people working in the room.

As a result of that evening and some subsequent meetings during my visit, we began a sporadic correspondence. It was mainly from India where he and Helen Spurway went to live. He seemed to enjoy living in India, although he could be as difficult with his new Indian friends as he was with the people he left. I asked him once why he had left Britain, and he said, looking at me as though I did not exist, "Because there are too many damned Americans here, especially damned American soldiers." There probably were, but I don't think that was the reason at all: I think there was quite a bit of the Hindu Brahmin in his nature, and he found some peace there for his turbulent mind.

I still have our letters, which span the years from 1959 to 1962. As I reread them I am impressed all over again with how fertile his mind was. He could look at any biological problem with fresh and ingenious insights. He was also not encumbered with a need to flatter. In the early 1960s I had sent him a book that I had written for the layman (*The Ideas of Biology*). Here are some fragments of his replies:

15 November 1960

Dear Bonner,

. . . You ask about Helen and me coming to Princeton. This is at present impossible for me. I was asked by your government to give a list of all associations to which I have belonged since my 16th birthday (in 1908) with date of joining and leaving, with a threat of jail or fine if I get one wrong. I don't know if I joined the

EVERYTHING PEAKS

Oxford University Liberal Club in 1912 or 1913. Having a professional regard for truth I am not going to guess. If President Kennedy has the guts to tear down this Iron Curtain I will come when next asked, if I can manage. But I think there are too many officials who have a vested interest in that sort of nonsense.

So you had better come here. There are plenty of molds, especially in the monsoon . . .

14 February 1962

Dear Bonner,

. . . Every 15 years or so I write a paranoiac paper. In 1919 I gave the dimensions of a gene, and several other things about genes, not wholly wrong, on very inadequate evidence. In 1928 I gave the general accepted theory of the anaerobic origin of life. I hope Nagy has bust it. In 1944 I produced a cosmological speculation which nobody likes, not even myself. Perhaps it is right . . .

25 September 1962

Dear Bonner,

Thank you for "The ideas of biology." I have not yet read it, but my first impression is that you have made a number of statements, sometimes for the first time, sufficiently clearly to allow destructive criticism. For example on p. 29 . . . [Then he makes five detailed points, all excellent, the last one concerning page 152—the book is only a little over 200 pages!]

Anyway the book is provocative, probably more so than you meant it to be . . .

I kept wondering what he might have said had he admitted to reading the book!

After giving those 1956 lectures, I had an invitation to spend the weekend with Victor Rothschild (well known for his work on fertilization), whom I knew slightly, and his family in Cambridge. It was the first time I had seen Cambridge and I was quite bowled over by its beauty. Both Victor and his wife Tess could not have been kinder and I had a wonderful time. I had previously sent the manuscript of my lectures to Cambridge University Press, and they had given me a very unencouraging reply. Victor asked about it, and I told him the details. He got up and said wait a bit, and disappeared into his study. I could hear him distantly on the telephone, and he came back to say that it was all settled: they would publish my book! Victor had not even read it—what could he have possibly said over the telephone and to whom? Of course, I never knew, but they did publish *The Evolution of Development*. Since then I have always wished that whenever I finish a book, there would be a Lord Rothschild about to speed it on its way.

Even my departure from London after those whirlwind two weeks was an event. On the night of departure, Ruth Lowinsky, the mother of Clare and Justin, who had stayed with us as refugee children during the war, invited me to a family dinner. She was a celebrated cook and that evening she outdid herself: the food and the wine were a dream. I kept worrying about catching the plane, as I always do, but Justin said he would drive me out in plenty of time. We started very late, and by the time we got there almost everyone had boarded. All the regular seats were filled, so they had to put me in first class. One of the other passengers was Maria Callas in her splendid elegance. The pretty stewardess came to me and said that having been upped to first class, I was to have a steak dinner with champagne. I explained to her that I had just come from a sumptuous dinner and could not do it. She was very upset because I was passing up this chance to have a fantastic meal—and free too! I rode home among the clouds.

Five years after my first leave in France I was again, in 1958, given a semester off and I needed to apply for another fellowship. It seemed to me the most sensible thing to do was to apply for a Guggenheim since I had begun my application there five years previously, but had taken the Rockefeller because it had come through first. In discussing this with my older colleague Elmer Butler, he urged me not to write them but to go in and talk directly with the president, Henry Allen Moe, who Elmer said was a person quite worth meeting. I made an appointment and went in to New York to see him. He was indeed a charming man, and I explained to him that I would again like to apply for a Guggenheim Fellowship, and asked what should I do. He said, with a great smile, I did not have to do anything: I already had the fellowship. I could not quite gather in what he was telling me, so I began to explain but he interrupted me, saying, "No, you're the one that doesn't understand—you are all set to go. You *have* a Guggenheim Fellowship." I finally realized that my previous application had gone through five years earlier and they were just waiting for me to apply again. I was rendered spluttering and speechless, much to the friendly amusement of Henry Allen Moe. I felt very lucky indeed.

I had wanted to go to the Genetics Laboratory of C. H.

Waddington at the University of Edinburgh. He was someone I did not know but I greatly admired his books and his research. He was a pioneer in that he was simultaneously concerned with development, genetics, and evolution, producing the very kind of synthesis between these three domains that I considered of the utmost importance. He made me feel immediately welcome in his department, and while I saw quite a bit of him, and always gained from our conversations, I cannot say that I really got to know him—he was not that kind of person. There was always a distant, British briskness about him that could hardly be described as warmth. Yet he had a good sense of humor. I remember near the beginning he took Ruth and me out to dinner at one of the fancier restaurants in Edinburgh. Our waiter was a young cockney who asked me what kind of potatoes I wanted with my sweetbreads. I apparently was slow in making up my mind, so the waiter said, "I would recommend mashed—with that sauce you need an absorbent potato." This was too much for Wad, as we called him, and he collapsed with laughter.

There were many other stimulating people in his department, so that I enjoyed myself thoroughly and learned many things from Wad and his colleagues. I had decided to do mainly laboratory work on slime molds, which I brought with me, and Wad took great interest in my progress.

When we arrived in January it was very cold. I noticed that everyone had an electric fire in their office except me. So I went to the stockroom manager and asked for one, and the man said something to the effect that he supposed that as a soft American I could not take this kind of indoor temperature. I quickly said he had it backwards: some Scot had stolen the heater from my room. We got along very well after that because I soon learned the Scots were kindly disposed toward Americans. We had two things going for us: they all had relatives in America, and we weren't English.

The laboratory temperature had another good effect. The university turned the heat off at noon on Saturday and back on Monday morning. It meant all my slime molds stopped in their tracks for the weekend and did not resume their activities until mid-morning on Monday. We all had the weekends off.

During that stay I began to worry why so many primitive organisms—and I had small, colonial algae particularly in mind—have an alternation of sexual and asexual generations. For instance, the colonial alga *Volvox*, which is a common inhabitant of our freshwater ponds, will have numerous asexual cycles that produce a series of identical colonies, or clones, and an occasional sexual cycle that produces a thick-walled, resistant stage. It occurred to me that this pattern had evolved through selection because in the benign summer when growth conditions are ideal, successful reproduction means producing as many offspring as quickly as possible. If the genetic constitution of a colony was suited and could thrive in those particular summer conditions, there was no need to vary—just duplicate oneself as rapidly as possible. In the fall, with the growing season coming to an end, and no certainty that those conditions would be the same the next spring, the safest way to ensure continuing reproductive success was to have a sexual cycle in which two genetically different individuals fused their egg and sperm, and the resulting offspring that germinated come the warmth of spring would be genetically diverse, and some of the variants would have an improved chance of coping with the new conditions. By this asexual-sexual alternation, *Volvox* could have it both ways: fast reproduction in constant, dependable growing conditions, and variable offspring to meet any change in the future environment.

The chair I sat in while writing this paper had a little plaque on the arm saying "Donated by L. C. Dunn." He was a distinguished geneticist at Columbia University and the editor of the *American Naturalist*, the very journal I intended to send the paper to. In my covering letter with the manuscript, I said that I had written the paper sitting in Dunn's chair and his spirit surely would make the paper acceptable to his journal. He sent me an amusing reply in which he said that when he had visited the Genetics Laboratory many years previously, the chairs were so uncomfortable that in a fury he went off and bought one. The then-head, Professor Crew, was much amused and had a plaque put on. My paper was published.

There was an eyeful of a young woman who washed everyone's dishes, and I soon heard that she had a reputation for being too friendly. I could see why, because she looked quite irresistible. Like every other man, I would linger for a chat when I was delivering or picking up my glassware, and we often discussed greyhound races, to which she was addicted. One day I arrived with a big basket of dishes, and before I had a chance to say anything, she fixed me with those siren eyes and said, "Let's go to the dogs together." I nearly dropped my whole basket but then I realized she was talking about something other than what I first thought, and recovered. I'm sorry to say I never went to the dogs with her!

Scotland always had and always will have a magic for me. It started in my early years when my family spent a few summers there, and it was strongly reinforced that year in Edinburgh. There is something about the brisk air, the green of the hills and the fields, the heather and the bracken, the burns and the rivers, the lapwings and the curlews, the skylarks and the robins that stirs me deeply. I even like the occasionally paralyzing cold, the soaking rain showers, and the changing sky. I have a special feeling for the small towns with the dark stone of their severe houses. Edinburgh is in particular full of marvels, from the ancient tenement houses to my favorite, the Georgian buildings of the New Town. The Laboratory was in the King's Buildings, which are right by Blackford Hill, so in a few minutes I could walk out onto a grassy moor full of gorse in bloom.

We were lucky and through friends landed a very pleasant house just a short walk from the Laboratory. It had been built a few years back by the philosopher Kemp Smith, who now was in a nursing home; he was happy to rent his house to someone from Princeton because he had fond memories of teaching there for a year. Being of the old school he had not put in central heating, so we had to plot to keep warm. For instance, I remember my electric razor became totally congealed and would not work in the morning. So the moment I woke up I would put the razor in the bed by the hot water bottle to get it ready for action. The remarkable thing was that the children did not seem to notice the cold and would play happily in a room that had no fire at all. Ruth and I would gather around the fire in the living room-library and would be joined by Professor Smith's black labrador, who had the amazing trick of parking his rear end on one of the low chairs by the fire with his front legs in front on the floor, and would then look right at me. I always suspected that is what he did with the Professor—they would sit and discuss philosophy together.

We made some very good friends on that visit in the Mitchison family, whose children were roughly the same age as ours. Murdoch was in the zoology department, where he later became the Professor, and Rowy was a historian who also became a Professor in the university. In those days a professor in Scotland was something far more impressive than anywhere else in the world, for the Scots revere education.

The Mitchisons loved to walk, as I did, and they introduced us to many splendid places that could be reached by short drives away from town. This was not so much on our first visit, but on the two subsequent leaves we took to Edinburgh and short trips in between. We could go to the Lammermuirs, where I could get my fill of heather, black-face sheep, grouse whirring off as one flushed them from the heather, hares racing away, and the soulful cry of curlews. Sometimes in the middle of the walk the sky would open, but when the shower was over one would steam a bit and then slowly dry out. And after many hours, there would be that wonderful feeling of fatigue in one's legs, to be rewarded by a pot of tea and some raisin scones and shortbread. Sometimes we would go down to the Lake District or Yorkshire to see the wild daffodils and walk the dales. There the reward was a delicious pub dinner, with splendid beer. I love walking anywhere, provided it is not too hot, but the open spaces of the moors and the dales cannot be beat.

That year I also did some fishing on the Tweed. I would

play hooky for an afternoon and go to the upper reaches where the trout and grayling were plentiful. The Tweed valley is one of the more beautiful places in this world: the river itself, the green fields hedged in, the rolling hills, and the low mountains in the distance. Scotland was where I had learned to fish as a boy, and now I was reaping the rewards. I saw hardly any other fishermen, and now and then I would have a fish. I will never forget the drives home, with the slanting yellow evening sun lighting up the hills studded with sheep. Some images never seem to dim in one's memory.

Murdoch Mitchison's father and mother had a big place on the west coast of Scotland, in Kintyre, and they invited us for a few days in June, when the rhododendron was in bloom. The drive there was spectacular and we could not have been made more welcome. Dick Mitchison was a member of Parliament and a man of great charm, and Naomi (Nou), his wife and the sister of J. B. S. Haldane, was a well-known novelist. From Dick I learned many things about Parliament—things that had puzzled me from reading Trollope. Nou had her desk behind the sofa in the bay window of a very large living room with a blazing fireplace and big windows that looked out toward the sea. She would blast away on her typewriter even when the room was full, oblivious of all the talk until she had finished. It is the kind of concentration I have always envied.

Kintyre is a wonderfully wild place, but the old house, with its turrets, its large Victorian windows, and its simple furniture, made one feel protected from the elements. More than that, there was a warm and lively atmosphere that came not only from the open fire in the living room but from our hosts and all the other guests. We must have been over twenty in all: other Mitchison young besides Murdoch and Rowy, Labour politicians, writers, artists, and a sprinkling of children. We all sat around an enormous dining room table, looking out on the blooming rhododendrons. Everyone served themselves at the sideboard and sat where they could find a place. On that first visit the cooking was done by Dick's political agent, who of course sat with us at the table, and there was massive cooperative help with the clearing up. It all seemed to run very smoothly and we had a marvelous time, as we did on numerous subsequent visits.

Some years after Dick had died we were there for New Year's, which is the big celebration in Scotland. Nou invited the whole village, as she did every year, and the whiskey flowed. There was a piper and a fiddler for the dancing, and the din was enormous. All the young practiced reels the afternoon before. Everything and everybody glowed that night, and the dancing was a perfect delight, especially watching some of the young mothers leading their small daughters. At midnight we all formed a ring and sang "Auld Lange Syne" as we danced back and forth. After a lot of kissing, there were two speeches: the first one by the senior fisherman thanking "Dear Naomi," which he kept repeating in a slightly sodden manner, and then Nou made her gracious reply, which had a moral to it but, alas, I don't remember what it was—all I remember was that it was just right. The next day we had to do "visits" to friends in the village. It was required that they offer us a dram, and it was required of us that we drink it. It was a grand time but we all needed a rest after it was over.

One Sunday we revisited Lennoxlove, the castle my

grandfather had rented when I was a child as a place to gather in the summer all the Swiss and American relatives. It was now owned by the Duke of Hamilton, and for a few shillings one could make a tour when the Duke was not "in residence." It was much as I remembered it from twenty-seven years before. I mentioned to the guide that I had lived there, and he clearly thought I was an American nut. It so happened I was wearing a tweed jacket my father had discarded and I remembered we had a photograph of him by the front door in that very jacket. So Ruth took a snapshot of me standing in the same place with the same jacket. The result was interesting: I was not wearing the long-gone plus fours that went with the jacket, and generally I looked rather like an intellectual bum; my father, however, looked like somebody straight out of Evelyn Waugh.

That first stay in Scotland in 1958 was the only time I was tempted to leave Princeton. One day Michael Swann, who was then the professor of zoology (and later vice principal of the university), came to see me at the house to tell me that the chair of botany was vacated and asked if I would apply. I was bowled over. The idea of spending the rest of our lives in Scotland was fundamentally appealing, not only to myself; Ruth and the children seemed to feel the same way. However, I had to give the idea some serious thought and begged Michael for time. I knew quite well I was just applying, and that there might be stronger applicants, yet here seemed a wonderful opportunity. I discussed the pros and cons endlessly with Murdoch and wrote my parents and colleagues in Princeton for advice. My parents said do it by all means, but the letters from Princeton showed gratifying horror. In fact, Princeton immediately promoted me to full professor, which made the decision all the more difficult, and I finally decided that I should go back to the United States. We had been happy at Princeton, it seemed a better bet for my work, and there were more unknowns in Edinburgh. Later I would now and then ask myself whether I had made the right decision and I always came to the conclusion that I had, simply because things worked out so well for me at Princeton. I became doubly convinced when Mrs. Thatcher started strangling British universities.

As eventful as the previous twenty years had been, those that followed were if anything more so. This was the era of the full blast of the civil rights movement, as well as the Vietnam War and the bombing of Cambodia, which fanned the student cultural rebellion. In biology not only was there amazing and rapid progress in molecular biology, but its applications to other branches of biology spread like wildfire. Evolutionary biology and ecology suddenly took on a new life and produced some major advances. As far as my life was concerned it was a period of tremendous activity, what with a growing family and administrative duties on top of my research and writing.

As for my research, there were two major bits of progress that I shall describe, one of which had an ancient history. Cap Weston would ask, in his sparkling lectures, how is it possible that minute fungi and slime molds have tiny fruiting bodies that invariably stick out at right angles from the substratum. They are not affected by gravity as are higher plants, which is understandable for they are too small for gravity to have much of an effect. It is presumably advantageous for them to stick up into the air, for this helps the dispersal of their spores, something important for their dissemination and survival. I did not realize it at the time, but my first clue as to what might be happening came from an observation I made as an undergraduate. If a migrating slug was cut into three segments and then allowed to fruit, the front and hind fruiting bodies leaned forward and backward, respectively. At the time the reason for this was totally misinterpreted by me. I said the front cells were faster cells, and no wonder they produced a fruiting body that leaned forward; the hind cells were laggards, so of course they leaned backwards! Ah, youth.

I took up the problem again twenty years later, in the 1960s. I am now going to describe some experiments in some detail to show the kind of experiments I enjoy most. My idea of happiness in the laboratory is to do experiments that somehow combine the utmost simplicity of design with the maximum significance of result. This has been my constant dream and no doubt that of all scientists. I think I came closest to that goal in my studies of gas orientation in slime molds, and for that reason I look back at them with affection.

The new attack began with two seniors doing their honors thesis work. They were able to show that one could rearrange the three separated segments of a slime mold slug so that the anterior or the posterior one was in the middle, and the end fruiting bodies still leaned away; it looked more and more reasonable to suspect that rising slugs were repelling one another in some interesting way. The following year my assistant and I did experiments with just two cell masses that we could place different distances from one another. The closer they were to each other, the farther away they leaned from one another as they rose into the air. By this time we suspected that the cell masses were giving off a gas that repelled, perhaps by making the cells on one side of the rising fruiting body move faster that the other, thereby making them lean away from one another. This could be tested in a number of ways: for instance, instead of using two fruiting bodies, we just used one placed near an agar jelly cliff in the culture dish, and the closer the cell mass was to the cliff, the farther away it leaned. If it was placed right at the base, in the crack, the fruiting body rose so it was exactly equidistant from the floor and the wall of the cliff. The crucial experiment came from placing a bit of activated charcoal near the cell mass, and then the fruiting body would develop right into the charcoal. Charcoal is well known for its ability to adsorb gases, so presumably what happened was that the repellent was removed on the charcoal side and there was more repellent on the other (back) side, which made the rising fruiting body move straight into the charcoal.

There was always the question of the chemical nature of the gas, and we did not resolve that until the 1980s. By placing a cell mass near a small hole connecting two compartments, we could put different gases on one side. A repellent would make the fruiting body grow away from the hole. Of all the gases we knew slime molds gave off, only ammonia was a repellent, and it was effective at extremely low concentrations. It took me forty years to be able to answer Cap Weston's question: small fruiting bodies jut out at right angles from the substratum by giving off a repellent gas, which if evenly distributed around their buds, will force them to grow straight up into the air.

Recently we have done some experiments on migrating slugs, which are so clever at orienting toward light and in heat gradients. We found that there is a possibility that light and heat affect the amount of ammonia produced on the appropriate side of the sensitive tip of the slug, and therefore the slug's orientation is the result of small local differences in ammonia production, which in turn affect the direction in which the slug moves.

This project has kept me occupied on and off for almost a

lifetime in the laboratory and has given me enormous pleasure. It has involved trying to solve a straightforward biological problem, and doing so by means of simple but revealing experiments. For me there is beauty in this kind of rudimentary science.

The other big event in my lab happened a couple of years later. Ever since 1959 we have spent the summers in Cape Breton, Nova Scotia, first in a cabin and then in a small house on the Margaree River. One morning in the summer of 1967 my work was interrupted by an excited telephone call from my laboratory in Princeton: David Barkley, a graduate student, and Theo Konijn, a visiting colleague from Holland, had made a big discovery. They were both on the line and they were talking so rapidly that I had to make them repeat everything twice so I could understand it.

As I have mentioned earlier, in my Ph.D. thesis research I showed that slime mold amoebae are undoubtedly attracted by a chemical substance when they come together to form a multicellular slug. For twenty years various workers had made attempts to find what was the chemical nature of the attractant, which I had called "acrasin." For a while it looked as though it might be a steroid hormone because a worker in the field had showed that urine of a pregnant woman had the ability to attract amoebae. Some years later a young assistant and I tried to repeat this finding using the urine from someone who was not pregnant, and that worked also. (The story was not quite that simple. She used a diluted sample of her own urine, and it worked. I pointed out to her that we had shown that the urine need not be from a pregnant person, and she blushed beet red, and said she had been meaning to tell me that she was going to have a baby. So we tried my urine and it worked too, and I was quite sure I was not pregnant.)

Theo Konijn had done some work that paralleled ours, and he got a grant to come to Princeton so that we could combine forces and attack the problem. He came shortly before we left for Canada to begin his year, and he and David Barkley were having some maple walnut ice cream discussing all that was known of the properties of acrasin. It was a small molecule, given off by bacteria, found in urine, and there were a number of other clues. David said that it had many of the properties of a new substance called cyclic AMP, isolated recently by Earl Sutherland at Vanderbilt University, which he had just learned about in a biochemistry course. Cyclic AMP was a key link for a cell between the receiving of a message from a hormone and informing molecules within the cell. For this reason Sutherland called it a second messenger; it was a major discovery for which he received the Nobel Prize.

So David and Theo obtained some cyclic AMP and quickly found that it was a strong amoeba attractant even at exceedingly low concentrations. For slime molds it was a primary messenger as well as a secondary one. That is what they were telling me on the telephone. I knew they had hit the jackpot, but I was at a big disadvantage, up in my northern isolation, because I had never heard of cyclic AMP! Their excitement was contagious—I could not continue my work; I had to talk to somebody, so I got in the car and went to see Ned Park.

Dr. Edwards A. Park was a perfectly remarkable man who had a big influence on me, as he had on many others, on how to deal with life. He was a pediatrician who for many years was the head of pediatrics at John Hopkins University. In early days he had significantly contributed to the studies that led E. V. McCollum to discover vitamin D, which was the dietary requirement to prevent rickets. He had been coming to the Margaree River for forty years, for he was addicted to salmon fishing. He lived near the river in a small complex of cabins that he had built some years before in a remote spot. When I went to see him with my cyclic AMP problems he was almost ninety years old and a widower. His much-admired and much-loved wife had died a few years previously.

He was over six feet tall and even in his old age as straight as a rod. His manner of speaking was old-fashioned in the best way. He had a strong but gentle voice; he spoke in a slow, deliberate manner. He had a wisp of a perpetual smile that lifted up the corners of his mouth and somehow gave one the feeling of warmth and understanding. He could see the humor of any situation. He told me once that when, as a young man, he first started to practice pediatrics at the New York Foundling Hospital before the First World War, kindly nuns did the nursing. One day he passed Sister Ann, a particularly sweet and unworldly person, in the hallway and she asked him what he was going to do Sunday. He said that if the weather was good he had set his heart on a long walk in the country. The next day she told him that Sunday was to be a sunny day. He asked her how she knew, and she told him that she had whispered the request into the ear of a dying newborn, and that it was well known that the totally innocent had direct access to the ear of God. Then Ned added, with that wonderful smile, "Now John, I want you to picture God sitting on his throne, surrounded by all his archangels, all grappling with the great problems of the world from wars to famine, and suddenly everything comes to a great halt as this little babe appears by his side and says, 'Dr. Park must have sunny weather this Sunday for his walk.' I might add, that of course it was sunny that Sunday."

It was those unexpected, unpredictable, and revealing quirks of behavior that delighted and amused him, as he would survey the human condition from his rather Olympian but sympathetic vantage point.

When I went to see him and told him about the extraordinary telephone conversation, and that I didn't even know what cyclic AMP was, Ned's eyes twinkled, and after offering some congratulations for my colleagues, he said not to worry, he could help me out. I could hardly believe my ears as he went on. He explained that he knew quite a bit about cyclic AMP and could get me reprints rather quickly because "Earl Sutherland has stayed with me in this cabin." It turned out Sutherland was in the department of physiology at Vanderbilt, which was chaired by Dr. Park's son, and they had both been up for some fishing. Indeed, I got all the information I could have desired in no time, and immediately was able to sound as though I knew it all along.

It has always seemed to me a bit of extraordinary luck that the solution to the chemical nature of acrasin should have come so easily. Theo Konijn came for the year to work on a problem that he solved in his first month. The result of this work was an enormous interest in many laboratories on how the cyclic AMP oriented cells, and today we have large amounts of biochemical and molecular information on the details.

Tenure at Princeton changed my life in few ways. The family was growing and the increase in salary was important. However, as far as my university work went, we all continued to teach roughly the same amount. I was now receiving outside research grants, which meant I had a laboratory assistant, and there were an increasing number of students doing research with me. All of that was a progressive, incremental process and had nothing to do with promotion to tenure. Eventually I went on to become full professor, and while the step pleased me very much for professional reasons, it also meant we were in a position to apply for a bigger university house, which we needed, for with the arrival of Andrew, we now had four children.

This apparently seamless progression came in for a big jolt in the spring of 1965. Bob Goheen, the president of the university, called up to say that he was coming down to see me right away. This put me in a fluster, but along with everybody else, I was a great fan of his, so I was glad to see him; but why was I not summoned to his office? The moment he came, he explained his mission: he wanted me to be the new chairman of the department. I instantly had two thoughts race through my head: I didn't want to do it, but I was going to say yes, both of which turned out to be true.

I was to take over the next academic year. My predecessor, Arthur Parpart, was about to leave for the summer for Woods Hole, but he promised he would tell me all when he returned, including the details of the finances of the department, which made me especially nervous because I knew my severe limitations as a bookkeeper. The awful thing was that poor Arthur had a massive heart attack just before he was to return and died instantly. After the funeral and the memorial service (which was most moving because there were no eulogies, just a Bach solo piece for the cello, played by a gifted colleague—music does things that words cannot begin to manage), I tried to gather in the reins.

The first attack came from the treasurer of the university, a splendid person named Ricardo Maestres, who told me that the biology department was paying the secretaries and the laboratory assistants too little, and that he could not let me see the department accounts. They were being fed into a computer for the first time and no one had them! I proceeded to make a whole series of wild guesses because Arthur had left nothing behind either—all the figures had been in his head. I fixed the salaries and flew blind all year. At the end of the year I had a meeting with Dick Maestres and all his accountants and associates. Dick liked these meetings to be solemn to make one feel one had done everything wrong, but what he told me was so outrageous that I came as close to blowing up as I am capable. He started the meeting by saying that I was paying the secretaries and the assistants too much, and I had been so economical that there was a surplus of money in the various accounts that the university would take back. The blood rushed to my vocal cords and I said, "You told me to do that for the secretaries, and your office helped me get the figures right. The only reason there is money in the accounts was that you still have all our figures stuck in your goddamn computer, so we had to deny ourselves many things we badly needed until we found out how much money we had." I realized a long time afterwards that Dick was just testing, and I must have passed, because we became good friends and I got what we wanted.

I learned many things that year that were revelations about human nature. When I started, an old friend who had the same job for a few years at the University of Edinburgh wrote to me that I should put up a sign on my office door saying "Departmental Chaplain." Slowly my distaste for the job began to recede a bit. It was partly because I was able to keep my research going without any serious lapse, and partly because there were a few small rewards in terms of things accomplished, things that worked. For instance, I had an older colleague whom I was very fond of, who had a mixture of aggression against the world (and especially authority) and a splendid sense of humor. One day he called me to say he wanted to see me about a matter that had been bothering him. I told him to stay where he was, and I would be right down. He started off in his aggressive vein, "John, you're the chairman, you're the authority here, and there is something that is your responsibility." When I asked what it was, he said, "People just aren't flushing the urinals in the building, and it's disgusting." I told him that would require some thought, and that I would come to see him the next day. When I returned, I told him I had the perfect solution: I was going to install a huge switch right over his desk that was connected to each urinal in the building, and whenever he felt the urge, he could pull the switch and they would all flush at once. There was a tense moment of silence, and then he burst out laughing. This shows why I occasionally had a feeling of accomplishment, but perhaps that was my greatest moment in all my years as chairman.

One of the biggest problems that faced biology at Princeton had to do with a change in biology itself, and what we were experiencing was a microcosm of a worldwide revolution. Biochemistry and molecular biology were making an explosive impact on biology. The advances in those twin subjects were so rapid and dramatic that the very fabric of biology was being altered in fundamental ways. The reductionist advances were giving new insights into the details of all biological processes. Many hailed these new developments as the important and exciting advances that they were, but many of the older biologists showed puzzlement at the incursion into the established ways. There were those who saw this as the way of the future, and in our department Newton Harvey was a strong advocate for our becoming more biochemical. For some years he had taught a course in biochemistry for undergraduates that was probably the first one of its kind in the country. However, there was no change until the early 1960s when the university finally decided to expand and set up a program in biochemistry that was to be a joint venture of the biology and chemistry departments.

It came into being about the time I became chairman. The first director of the program was Arthur Pardee, a first-rate scientist and a fine, decisive person. He and I came to be good friends, and as he recruited new people, there seemed to be the genuine possibility of harmonious relations. My own work with slime molds increasingly bordered on biochemistry and he became my guide and adviser; we even had joint students who worked on biochemical aspects of slime mold development. It was an exhilarating period for all of us.

This era was a fascinating one from the point of view of the sociology of science. Biochemists and molecular biologists did not infiltrate quietly nor as missionaries; rather, they saw themselves as the ones who would provide all the answers, and as a tribe were exceedingly assertive about it. They had found the new Truth, and all the rest of biology was fossilized dry rot. This amazingly aggressive attitude has only in recent years shrunk to more normal proportions, although it has not disappeared completely. A young molecular biologist today will still alarm me by saying that evolutionary biology is not science, not the real hard science; that is only to be found in studying molecules, either of the genes or of cells. Somehow the great success in these new fields led to an "I'm king of the mountain" syndrome that was conspicuous among the great and the lowly. This comes through vividly in James Watson's book *The Double Helix*; it is a great book partly because it describes that "take no prisoners" attitude so thoroughly. It was an attitude that caused consternation and chaos in many institutions all over the world, and certainly Princeton University was no exception. The very same thing happened at Harvard, and is admirably described by E. O. Wilson in his autobiography, *Naturalist.* I have often wondered what caused the warlike attitude and the extraordinary lack of civility, but I should probably leave it to the historians of science to get at the root of the matter. I have wondered, for instance, if it could be the influence of Max Delbrück, a pioneer in the field, who used his great intelligence, wrapped in his distinguished background in theoretical physics, as a battering ram on the sensibilities of lesser mortals. But then his co-pioneer, Salvador Luria, was a man of great charm—why was he not imitated, even by his own student James Watson, whose notorious brusqueness is so clearly Delbrückian? Perhaps that arrogance had less to do with individual people and more to do with their revolution; such attitudes were the trappings of those who wanted to replace the old order, and felt the only way to make a clean break for the future was to shoot the past.

Being chairman, I was in the thick of it. To begin with, there were administrative turf fights, which always boil down to fights for money, and there were intellectual fights concerning the future of biology. In my naiveté I had not realized that the rest of biology needed defending; I knew we had some extraordinarily gifted nonmolecular biologists in the department, and I just assumed that since the outside world recognized that fact, there was no need to convince our own administration. How wrong I was. I found out much later that the deans and other university brass were being bombarded with the news that all nonmolecular biology was in its death throes, and the only future was molecular. I found this out from a good friend, and an exceedingly gifted biochemist-cell biologist, who was leaving Princeton and has gone on to achieve great things. He had a parting conversation with the president, whom he told that the university must not neglect the other sides of biology because that is what Princeton was becoming famous for. This was surprising news to the powers; they were under the impression we were slowly expiring.

This revelation was not exactly the end of our troubles, but it was the turning point. It is true that with years one tends to forget the bad things in life and remember the more cheerful ones, but I cannot completely erase from my mind the time utterly wasted on those endless struggles for space, for teaching assistants, for secretaries, and for things of far more trivial significance. Despite the constant pushes and pulls I was able to prevent the demise of the rest of biology at Princeton. Some of my colleagues thought I was not confrontational enough, but one day one of them who was trying to get me to agree to his plan said to me, with some irritation, "John, everyone thinks you're a nice guy, but I've noticed that you always end up getting your way."

One spring day during an exceptionally busy period I was driving through the countryside near Princeton and passed a farm that sold grass sod for making instant lawns. In the middle of a beautiful sweep of grass was a big sign saying CULTIVATED SOD. I suddenly had an insane desire to come out in the middle of the night, steal it, and arrange it outside the door of the chairman's office. Of course, I never did it, but somehow the thought greatly helped relieve some of the tension that was rising in me.

It was during this period that Princeton became coeducational. It was another indication that the world really was changing what with the civil rights movement and the rise of feminism, and now Princeton followed suit with its own step forward. This met with furious resistance from some of the older alumni, and I always felt sorry for the first women to arrive, for the boys had a confused idea of how to behave. I suddenly had women in my basic biology course, and the only effect I can remember was in the beginning going into the men's room before the lecture to be sure my fly was zipped. It took no more than two years for all of us to find it difficult to believe that we had once been all-male—everything seemed so normal now that we had what was quaintly called "sex-blind admission."

When a women studies program started I was appointed to its committee. It was a very interesting, and indeed an eye-opening, experience for me. Furthermore, I felt quite useful because I plugged hard for having some biology taught besides the history, the literature, and the other relevant disciplines. The newly appointed director was an anthropologist and she embraced the notion with enthusiasm. This is one of the reasons Princeton had a strong program right from the beginning. My only problem was keeping the distinction clear in my mind between the meanings of gender and sex—it was rather like trying to remember the meaning of dialectical materialism.

I continued to enjoy my teaching. Lecturing on general biology to a large class was something that suited me. The students were responsive and I seemed to keep control of that sea of faces, almost always friendly and attentive. Sometimes they would needle me; I remember one young woman who sat in the middle of the room, and whenever my eye would fall on her she instantly produced a large pink balloon with her bubble gum. Her timing was perfect and it always came as a surprise and produced a glitch in whatever I was saying. For forty years I taught 150 to 200 students each year. I have often been greeted by what seemed a total stranger with "You won't remember me, but I took your course some years ago." They never have seemed angry with me but often remembered some of the rosy things.

Teaching research to graduate students and seniors was

also a great source of enjoyment for me—I always was trying to follow the footsteps of Cap Weston. The rewards were the pleasant companionship, the feeling that one could really help when there were problems with the experiments (or with the cultures—I could have started a business as a slime mold veterinarian!). The hardest thing to instill into the very bright students was the idea that one can never predict how an experiment will come out. I had a graduate student who would come to me with a very good idea of something to try. I would urge him on, and the next day when asked how it was going he would reply that upon thinking about it some more he realized it would not work. I would explain he must try it and get the slime mold to speak to him, but it was of no avail and he took an instructor job at a college without his Ph.D. A few years later he submitted an excellent thesis and I asked him how he had managed to loosen up, and he confessed he finally understood what I was telling him when he had to oversee undergraduate research; teaching had taught him. Over the years I found the very same thing to be true for myself: as everyone knows, the best way to learn is to teach.

The 1960s and 1970s were an extraordinary time of change for young people. Largely because of the Vietnam War—although perhaps that was just a trigger that released something that was going to happen in any event—the young staged an impressive revolution. The students at the university, as did the students everywhere, began to assert themselves. It was suddenly as though there was an instant formation of a powerful union, and that union was an irresistible force that could impose its will on the older generations, which were in shock and disarray. The outward signs were most obvious in the male students: no neckties and long hair. And for both sexes, the universal blue jeans, free use of four-letter words, the disappearance of chastity, and the appearance of drugs. At one point the students demonstrated around a building on the campus where secret military work was being carried out. A next-door neighbor told me that at dawn, before the crowd gathered, a small, undeveloped young girl of about fourteen was painting "Fuck the Bourgeoisie" on the wall of the building.

At the university requests became demands, and the radical element among the students took over the undergraduate newspaper and would attack the university officials and faculty—any authority—in the most virulent fashion, a virulence that has only been matched by the far right in recent years. The barricades were finally stormed when our government bombed Cambodia—that was the last straw (although there were quite a few last straws, such as the killing of the students who were demonstrating at Kent State University). The outcry was deafening and the whole university became convulsed. Besides student demonstrations there was a series of marathon faculty meetings that were wisely broadcast on the student radio station so that all could know that we were not conspiring (and the spouses could follow the drama and know when we were coming home for dinner). Bob Goheen, our president, showed remarkable decency, patience, and enviable steel: his role was crucial in averting any serious trouble or any regrettable turns.

Two of the undergraduates in the biology department came to see me to tell me that they had called a meeting of all the students, graduate and undergraduate, all the laboratory assistants, secretaries, janitors, and department faculty for the next morning. I told them they could not have it then; that would interfere with classes. They informed me very politely they were quite aware of that, and they hoped I would come! Of course we all did, and our big lecture hall was absolutely jammed—I do not think there was a soul anywhere else in the building. One of the students started the meeting, saying that it was held because of the outrage of bombing Cambodia, and it was time to discuss what was happening to the world. After a few minutes of a rather tense beginning, the convener called a few fellow students to the podium and there was considerable whispering. The leader then looked straight at me in the back where I was sitting, and said they wanted me to run the meeting—they were not used to doing that sort of thing. I was simultaneously surprised and flattered by their confidence, but it was an easy meeting to chair. To a woman and to a man, we all agreed that the war was misguided and, worse, immoral; it had to be put to a stop. We passed a resolution that took the form of a telegram that I was to send immediately to President Nixon, but leading up to this resolution was a remarkably good discussion not only of the war that outraged us all but other matters concerning student needs and frustrations. It is the only time in the long history of the biology department where we had everybody, at all levels, meet in the same room. Catastrophe and the emotion that goes with it brought us together in a way that untroubled times never do. It was a moment we shall all remember for we were grateful to be unified. It is a pity that only something quite dreadful can do this, but then it was a soothing antidote to a horror we were all feeling.

Some of the demands students were making were not

realistic. In some departments there was the demand that at the beginning of a course the students meet and decide what they would be taught. I received such a delegation and was told that was what we had to do in biology. I urged them to think about what they were saying for a moment: how could they decide what should be taught before they knew enough biology to make a sensible decision? I said, "Look at a more extreme case: how could students in an atomic physics course provide any kind of a sensible outline to a faculty member teaching the course before the students had learned some atomic physics?" Fortunately they saw my point.

Some time after that the premedical students came to me and said that we should abolish grades; people should be captured by the subject and not grub for grades. I told them that in principle I could not agree more, but there were some problems with such a scheme. Since they were about to go home for the Christmas holiday, I wanted each of them to get in touch with the admission deans of their local medical school and ask them their views. I suppose it was a mean trick on my part because when they came to see me after the holiday they all reported that all the deans had also agreed with the virtue of the principle, but added that the student should not bother to apply to their medical school. It would be impossible to evaluate the student, and they had so many good applicants that they would have to choose among those who had grades. Again a commendable idealism gets blown to bits by reality. The good part of all these ideas, even the abortive ones, is that they changed the university very much for the better. Today students continue to be able to have a say in many things that affect them, something that was not possible before the revolution.

During this period our own children went from Rebecca who was in her early twenties to Andrew who was in his beginning teens. In one way or another they were all affected by the revolution. Rebecca got herself arrested in a civil rights demonstration trying to confront Lyndon Johnson; Jonathan became a flower child and starved himself so he could avoid the draft by being underweight; Jeremy quietly let his hair grow longer; only Andrew was too young for any visible manifestation of the period. Ruth and I did not find it easy to understand, although we slowly became educated. Oddities in dress and behavior seemed the norm for other people's children, but it took on a different meaning when they were one's own.

One day our gentle Jeremy and I were towing a dying car to the garage, and what we did not know was that it was illegal to tow with a rope—it had to be a stiff rod. It was the early days of the cultural changes and Jeremy looked exceedingly hippie with a beard and a pony tail. As the policemen who stopped us got out of their car, Jeremy said to me in a soft voice, "You'd better let me handle this, Dad." Stunned, I acquiesced with considerable misgivings, but the policemen just politely told Jeremy not to do it next time. I was the one who did not have it all worked out. Ruth adjusted much faster, but we both made it in the end, and what is much more important, the children survived without a visible scar.

I can remember a student coming to see me during that tempestuous period and asking me if I would answer a few questions for his sociology senior thesis. I agreed, and he asked me my attitude on all sorts of current issues concerning his generation. I answered as best I could, and he said he could not understand it; my answers were quite different from the previous faculty members he had interviewed—I seemed to be much more broad-minded. I asked him if the others had teenage children and he confessed he never thought of asking. If one is not learning from students, one is learning from one's own children.

Returning to biology, having spent some time on the rise of molecular biology, I would now like to turn to the equally huge advances in evolutionary biology and ecology. They involved less aggression than accompanied molecular biology's great leap forward, but their effects on our thought today have certainly been of comparable importance. Without doubt the most significant advance was that of William Hamilton, who, from his studies on social insects, realized that the degree to which individual organisms were related affected the degree they would help or compete with one another. This came to be called "kin selection," and it became one of the particularly important explanations as to why some individuals cooperate while others compete. The greater number of genes two organisms shared, the more likely they would help, that is, be altruistic to one another. This led Richard Dawkins to write a brilliant popular account in *The Selfish Gene*, in which he extended and underlined Hamilton's argument by saying that the genes themselves were the ultimate units of natural selection, and that kin selection was a way of saving successful genes.

From this moment in the history of biology there arose an important discussion that started as rather an angry debate. Dawkins was vigorously attacked by some for his idea that gene selection was everything; it was a totally reductionist position, a position that met with disfavor by many. To begin with, everyone knew it was organisms, as individuals, that were selected; this was the position of Darwin and it remains our position today. Nothing has changed that basic view, but more than anyone Dawkins made us think of a bigger picture. Dawkins's position also ran into disfavor among dedicated Marxists because of their doctrinaire rejection of any kind of bald reduction; their view was a holistic one with the interactions between the parts being the way to look at biological phenomena. The important idea, started by Dawkins, that finally emerged from the fog of the debate was that selection can occur at more than one level: genes, the individual, and even groups of individuals. This is the pluralistic view that emerged with time. Today we appreciate that selection can occur at all levels, including among the organelles that coinhabit within the cell. Furthermore, cells can compete with one another within a multicellular organism; then there is the traditional view that the multicellular individual is a unit of selection. And as those individuals make up a colony, as in social insects, the colony becomes a unit of selection. Each level of organization that arises during the course of evolution, events that today we term "major transitions," produces a new level of selection. Selection operates at all levels during the course of evolution; everyone that took sides in the early debate was right. Note that all these Hamilton-derived ideas are another instance where the fusion of genetics with another discipline, in this case a second encounter with evolutionary biology, has produced sparks that have led to forward progress.

A related important event was the publication by E. O. Wilson of his *Sociobiology* in 1975. This huge book, beautifully illustrated, had an enormous impact because, among many other things, it explained and illustrated Hamilton's idea in a way that was accessible to all. Furthermore, it brought together not just social insects but all the social grouping of all animals, even down to slime molds! I wrote a very favorable review of it in *Scientific American* and remain just as enthusiastic to this day. Unfortunately one of the reasons it made such a big splash was a vitriolic attack by a group of prominent Marxists, which undoubtedly boosted its sales. Their claim was that the book promulgated genetic determinism of behavior, which is a Marxist sin. It seems today to be a rather antique point as we learn so much more about the genetics of behavior. Doctrine is not always a good guide for the future. Ed Wilson had to suffer during that period, but he prevailed and went on to even greater heights.

I think it is significant that both Hamilton and Wilson were students of social insects. It gave them, and is giving others, such as Mary Jane West-Eberhard, a vision that sees all of biology in a deep way. Perhaps even the lowly social amoebae—the slime molds—can open one's eyes.

Soon after I began my administrative duties in 1965, Robert MacArthur expressed interest in switching to Princeton from the University of Pennsylvania. We wanted to begin a larger program in ecology and evolution, and he seemed like the ideal person to get it started. At the time he was in his thirties and already was considered the leader of the new wave of interest in an ancient subject. What he had done, before he came, was to show that through the use of mathematical models one could get simplifying insights into the great complexity of nature. This was something first championed by his thesis adviser at Yale, G. Evelyn Hutchinson, but Robert went much further with it. The community of ecologists at the time was very opposed to his work. Any kind of simplification was a travesty because the magic of nature was its complexity, and any attempt to change this, especially one using incomprehensible mathematics, was a blasphemy. I think Robert rather enjoyed his role as a counterculture figure. He was the first to say his mathematical models would be replaced by better ones, but even his earliest attempts shed a large amount of new light. The models were simplifications, but as he said in his own words, "Where would physics be without frictionless pulleys?"

Immediately some new appointments were made, one of which was Henry Horn, an exceptionally imaginative and interesting person whose work on the geometry of trees became a classic. There was also a sudden influx of first-rate graduate students, postdoctoral fellows, and visiting scholars. Almost overnight Princeton became a Mecca for modern ecology and evolutionary studies. I remember with some amusement the tremendous scorn from the biochemists— all ecology was nature study and soft science, not real science. There was consternation when Robert was elected a member of the National Academy of Sciences, and the biochemists decided to take another look. They asked him to give them a seminar to explain what it was all about. Naturally he did not convince anyone, and he received a very bristly set of questions at the end. Instead of getting mad, he seemed to revel in it. Finally someone asked him why he thought any of the things he was working on were worthwhile. He smiled and said that was rather like asking someone why they thought the music Beethoven wrote was worth listening to.

His first office was right down the hall from my laboratory, and his first house was a hundred yards from ours, so inevitably we began to see quite a bit of one another. I liked him from the start, both for his crystal mind and his inner strength. He was very reserved and never said how he felt, and somehow he made one feel one should not say how one felt: those were things that should be understood, not expressed. He was a man of few words, and when he did speak he had a slight impediment, a catch in the beginning of his sentences. If he approved of something one had done or written, he never oozed diplomatic praise—he would let you know in quite indirect ways. If something was wrong he would tell you immediately. We also had quite a bit of communication on family matters because he and his wife Betsy also had four children.

What was especially important to me was learning about his ideas, which were quite new to me at the time. He asked me to read manuscripts he was writing, and I remember going over an early draft of the manuscript of a now famous book he was doing with E. O. Wilson to be called *Island Biogeography*. It made a big impression on me. I think I was even able to help with the book because Robert had a tendency to say things with as few words as possible, no doubt because of his mathematical training, and that did not always make for easy reading. I was a help because I was able to put my finger right on the things I did not understand, all of which were fixed with a sprinkling of added words.

Our unstated friendship grew and we decided we were too cooped up, and since we both liked to walk, we should play hooky once a week and go for a good walk (with binoculars). We settled for Thursday afternoons, and at the stroke of noon we would rush off in a car with sandwiches in our pockets. We went to a variety of places: the sand dunes of Island Beach, trails in the woods, flat miles along the Raritan Canal, sandy roads in the Pine Barrens. He was a gifted field biologist and nothing went unnoticed. He could identify a bird from minimal cues, and he was always right when we got a better view.

I am not sure how aware I was of it then, but it is plain to me now that it was a period of expansion of my horizons. I was always interested in big questions, and especially how different parts of biology fit together, but I was ignorant about many key things, especially those where Robert was a master. It suddenly was possible for me to see that ecological communities were not just a hopeless tangle of charming natural history but a beautiful edifice that linked together so many of the things I had been thinking about. It was during that period that I began to appreciate more clearly how everything was connected: development (my own subject), animal and plant physiology, community ecology, and even animal behavior. I began to see that the unity was because all organisms are life cycles, and that evolution by natural selection is what controls the nature of those cycles. I was entranced by my long discussions with Robert on our walks and was profoundly affected by them. It is not so much that I saw anything totally new, although there were many new pieces: it was more that it all became more organized and shifted into sharper focus. The living world was made of life cycles, each one of which involved genes giving off their signals, their instructions, that were carried out by a whole cascade of living processes: development, the functioning of the organisms, that is, their physiology, and in the case of animals, their behavior. Those walks were an important part of my life.

At six o'clock one Sunday morning the telephone shot me out of bed. It was from Alan, one of the MacArthurs' sons, who said his dad told him to call, that he was in tremendous pain, and would I come right away. I jumped into my trousers and was there in no time. I knew Betsy was away visiting her mother. Robert had an excruciating pain—it seemed to be in his kidney, and I assumed it was a kidney stone. I called for the ambulance and bundled him off, and reached Betsy by telephone. By the next day Betsy told me it was cancer of the kidney, which had spread everywhere. He was told he had about a year to live, and his bad kidney was removed immediately.

It was a chilling moment, but Robert pulled us all through. Right after the operation he was completely in charge of himself, would force himself, with his hand on my shoulder, to walk up and down the hospital hall to build up his strength as quickly as possible, and saw to it that no one talked nonsense to him. He told me when we were alone that this sudden change in his life was a huge shock, but that was the way it was—it had to be accepted. His recovery was quick, and he did not seem different in any way for some time. He decided that he would write a book summarizing his ideas.

As he began to fail our Thursday afternoon walks would become increasingly modest, with frequent stops while he would sit on a log to catch his breath. Eventually we would make our outings in the car and park in some lovely spot to admire the colors of the turning leaves. It became hard for him even to get into the car, but he loved the sights of fall, and sitting in the car munching sandwiches, our conversation lost none of its spark. He knew, one dreary week in early November, that I was going to take our youngest, Andrew, on a college tour. I went to see him the evening I got back. He wanted to know all about the trip for he was very concerned about the education of his growing children. There was a special feeling about that short visit that I did not understand until early the next morning. Betsy called to say that he had died during the night.

He was an intellectual prince at a very early age, and by the time of his premature death in 1972 his impact had been enormous. His view of nature had become the new Establishment—in a very short time he had created a revolution. A few years after his death I remember talking one evening to a young and exceedingly bright (and successful) British molecular biologist. He asked me if I had known MacArthur, and I explained how we had come to be friends. He said that he had been a senior at the University of York when Robert died, and felt so emotional about it that he almost became an ecologist to carry the torch. It was only after he calmed down that he understood that mathematical ecology was not his natural bent.

Robert was able to finish his book before his death; it was an important work that made a fitting envoi.

In 1963 I was granted another leave. I had been so struck by my short visit to Cambridge that I decided that would be the ideal place to go. Furthermore, I had my old friend Brian

Shaffer there—it would be good to be with him again, even though I did not plan to work on slime molds.

I wanted to write a book about organisms being life cycles. I was becoming increasingly convinced, as I thought about the big problems of biology, that this was the way to bring the concepts of embryonic development and evolution together. What evolved were life cycles; the entire cycle varied and changed through natural selection. Unicellular organisms were small with a correspondingly short life cycle, while a large tree or an elephant was not only enormous by comparison but had a greatly lengthened cycle from egg to adult. A major part of the life cycle that changed in an especially significant way during the course of evolution was its development.

As I look back on these successive books on development and evolution, I realize they say something about the progression of my own life cycle. Each one reflects the next stage of my teasing out the problem, but the central theme of organisms as life cycles has not changed. As I wrote over the years I began to see things with increasing clarity and found better ways of expressing those ideas. During the process, which has spanned my whole career in biology, I was never conscious of the continuity of my thoughts or that they were evolving. I just had them in a sort of naive, blind way. Sometimes I would get all excited because I had found a new way of looking at something, but then the next day I would realize that the reason I liked the idea was because I already had it a few years previously, and sometimes had even published it! In any event the context of my thoughts all had the same basic frame, although what I had to say within that frame was a continuous accretion. The book I planned to write in Cambridge turned out to be a step in that process.

I became increasingly aware of the importance of size in influencing the nature of life cycles. The bigger the plant or the animal, the longer the period of development, the longer the entire life cycle, and the fewer generations for a given interval of time. Size and its corresponding life cycle had enormous ecological consequences, for the environment consists of a great gamut of size niches spreading from the smallest microorganisms to great trees and giant animals of all sorts. Furthermore, size is deeply correlated with behavior in many ways that we easily recognize; for instance, everyone knows that the larger fish eat the smaller fish and so ad infinitum. Size has also many physiological interrelations, which in turn fit in with the behavioral and ecological aspects of life cycles. To give an example, the larger a motile organism, the faster it can move, be it a swimmer, a runner, or a flier. I tried to bring all these interlocking elements together in a book that I called *Size and Cycle*. It is telling that when the book came out in 1965, while I was pleased with its reception, I felt quite uneasy about it; I felt it was unfinished and that while many things fell into place there were still many that did not. It was an interim report—more thought was needed. I was still in the growth phase of my own life cycle.

Brian Shaffer found us a lovely, ancient house in Sawston. It always amused me that this is the locale of E. M. Forster's novel *Where Angels Fear to Tread*. What an appropriate place to try and produce some radical thoughts! We arrived in the beginning of February and that was the coldest winter

Britain had suffered since the seventeenth century. The house was heated by a few coal fires here and there (each using a different kind of coal) and all the water pipes were exposed. Some had burst before we got there. To top all that, the trunk with our warm clothes did not arrive until two weeks after we did. As usual the children did not seem to mind, or even notice, the cold, but Ruth and I had a hard time adjusting. At the desk in the study where I wrote I would have to wrap a blanket around my feet, but even with that I ended up with an enormous chilblain on the back of one heel. At night we would often decide that the reason for opening the windows wide was that it was slightly warmer outside than inside. We would go to bed with sweaters, scarves, woolen hats, and gloves, the latter making turning pages of one's book very difficult. (What we needed were gloves with a slit for the trigger finger.)

The two younger boys went to the local school. After the first few days Ruth got a note from the teacher saying we were not to send them in long corduroy trousers (what the teacher called "bags") but in flannel shorts: bags were unhealthy. Jonathan, our oldest son, went to a private school, where they taught all the courses he needed for our return to Princeton, except biology. It was my job to tutor him from a textbook that his high school in Princeton had given him. This was a big test for our relationship. Periodically he would tell me I was wrong, and then I would find the place in the book and he would concede the point. One day, after many of these mini-arguments, he finally said to me, with a surprised look, "Say, Dad, you know quite a bit about biology." A great moment in a parent's life.

I also spent some time working in the library of the

zoology department and would join the faculty for tea. It was an interesting experience for it was so different from the University of Edinburgh. The degree to which the Cambridge faculty made a point of appearing aloof and detached was quite amazing. Edinburgh had been much more like an American university in that regard. I got to know some of the Cambridge biologists quite well, and they were fundamentally no different from the rest of mankind, but that did not mean one said "good morning" if one passed in the hallway; one was supposed to look right through one another. I would purposely give friends the big hello just for the shock effect. It was different at tea—that is where I got to know people. I frequently sat next to Professor Vincent Wigglesworth who had been such an extraordinary pioneer in the study of insect development and physiology. We often discussed experiments, and one day we were talking about something I had done with slime molds that had not worked. I said my trouble was that I simply could not think like an amoeba. He thought for a while and finally said, "You know, I believe I can think like an insect." Considering what he had done, I knew it must be true.

Brian Shaffer had also managed it so that I was a temporary member of the Senior Combination Room of Gonville and Caius College, which meant I could go to dinner any night I wished. That was an interesting experience all in itself. There was pomp and ceremony, good food, and very good wine. There I saw quite a bit of Joseph Needham, which I much enjoyed. He was initially an embryologist with an encyclopedist's bent, and later became famous for his monumental history of Chinese science, a prodigious achievement. I remember one day he brought a guest who was a great authority on the history of the brush, and in one evening I learned far more about the brush than I ever wanted to know. All through his guest's monologue there was a twinkle in Needham's eye as he would look about at us in the room.

The biggest event came when Brian invited me to a feast. One had to have an appetite the size of a starved boa constrictor to cope with the fantastic number of very good courses. Each had its own excellent wine, so that everyone was quite "jolly." What was especially impressive was the appearance of the entire festive occasion. The students were all in black gowns, the dons in crimson robes. Everything was lit by candlelight. The next day the Master of the College asked me what I thought about it and did I have a good time. I said I had a wonderful time, and that it gave me the lovely feeling that I had suddenly walked into the eighteenth century. Perhaps because he considered himself a modern scientist, he was quite offended, which had not been my intention at all.

The subsequent leaves in 1971, 1978, and 1984 were all to the zoology department at the University of Edinburgh. During each of these writing was the main occupation; I was always trying to extend my ideas on development and evolution. I became fascinated again by the similarities between behavior and development. Both seemed to be in some way divorced from processes that involved strict gene control. In the case of development, all the cells had the same genes, yet during development those genetically identical cells produced a great variety of different kinds of cells; in an animal there were nerve, muscle, blood, liver cells, and many more.

In behavior, while some actions may be directly programmed by the genetic constitution, others can be learned and therefore are quite removed from the genes.

For instance, while all birds inherit some sort of basic song, in some species that song is greatly modified and perfected by learning from hearing the song of older birds. The genes have produced the wiring that is responsible for the basic song, a process that in itself must involve many steps from the first gene-produced proteins to the laying down of the complex pattern of nerves in the brain that are responsible for the song. And then there must be an additional set of genes that are the basis of learning, which permit the song to be further modified. So the edifices built during development and the behavior of animals have this interesting common denominator: there are many steps between the gene and the end effects. Perhaps the lessons gained from one will give us insight into the mechanisms of the other. That was my prime motivation for delving further into the problem.

In the late 1970s I decided to examine the evolution of behavior, how it removed itself from the strict restraints of genetic control, and this led to *The Evolution of Culture in Animals*, where I trace the process of behavioral emancipation from the genes right back into primitive multicellular organisms. It has all to do with my nonstop interest in what happens in behavior and development after the initial expression of the genes; they fire the starting gun, and then all sorts of interesting things follow, sometimes after many removes. It is those "interesting things" that are the future of modern biology.

The book was well received except among some anthropologists. I had defined "culture" as the passing of information from one individual animal (human or otherwise) to another, and they felt that as their word "culture" had a very special meaning for human beings and all the things that are associated with human societies. It was clear I had no right mucking up one of their private words and trampling on their turf. On the other hand, there were some biologically minded anthropologists who welcomed my book. I soon realized that I had unknowingly stepped into the middle of a hot dispute between two camps of anthropologists who were shooting back and forth at one another, and I was standing in no-man's-land.

I have been neglecting one aspect of my life that has been very important to me and spans both this twenty-year period and the previous one. It has to do with my love of fishing and spending the summers in the Margaree valley in Cape Breton in northern Nova Scotia. When we first came to the valley in 1959 (on a chance rec-ommendation of a friend), all but the main roads were dirt, and the farming was done almost entirely with horses. The first few years we had a cabin on a farm, and it was bliss for the children. They could help bringing in the hay, and Jonathan would lead King, a huge workhorse, away from the barn, to lift the loose hay up into the attic loft with the pitching machine. There were fewer bridges, the small ones single lane, and more ferries. One night I went across the island to pick up friends at the Sydney airport and when we got to the Bras d'Or, an inland sea loch that cuts down the middle of Cape Breton, at one in the morning, we could see the ferry on the other side. There was a big sign saying that to call the ferry after midnight we should press the bell in the middle of the sign. It was pouring rain, so we dashed out and pressed it. Nothing happened, so we went out and pressed again. Considerably later we took a careful look at the bell from behind the sign and could see that there were no wires attached to it at all—it was just there to make one feel good. After about two hours the ferry came and we were rescued.

Another sharp memory were the pit ponies. The local coal mines were not yet totally mechanized and they used ponies for moving the coal inside the mines. The month of August was a holiday for all the miners, and they brought the ponies up to the surface to gambol in the fields and eat green grass for the month—the rest of their year was spent in eternal darkness.

It all changed rather rapidly in the 1960s. Most of the roads became blacktop, and tractors replaced the horses. During the same period my brother Tony was experiencing an almost identical evolution in his small village in Mallorca. He told me that every spring on a particular Sunday (a patron saint of animals' day), the farmers would bring their beasts (mostly mules) to the courtyard in front of the church to be blessed by the priest. The first year of tractors, a farmer brought his tractor. In church doctrine animals and machines are distinct, so this practice was discouraged. In many ways something important was lost by those changes, but the hard life of a farmer was made easier.

As our youngest, Andrew, grew up enough to be left with his siblings, Ruth began to join me on the afternoon fishing expeditions. She soon developed into a first-rate fisher and cast an elegant fly. She concentrated on trout, which became her passion, but inevitably she hooked a salmon. It was a memorable occasion. We saw a fish stirring in the pool nearest our cabin and I urged Ruth to give it a try. She had on hip boots and could not wade out far enough to reach the fish, so we traded and she put on my high waders. She was lost inside them (and my feet were horribly compressed in her boots), but she cast out to the right spot and almost immediately was on to a big fish. She complained it was too big and wanted to give me the rod, but I urged her on and soon she was in the middle of a most dramatic struggle with a very active fish. Because she was very unsteady in the oversized waders I grabbed them from behind, and together we followed the fish. By that time my father, who was visiting, appeared, and he became as excited as the two of us. He kept calling encouraging things to Ruth with gentle advice, and then would roar at me things that always started, "For God's sake, John . . ." get her to do this or that as though Ruth were not there. What I did not realize was that an entire road crew of men had left their machines to gather by the river's edge to watch the fun. After a great struggle the fish was finally landed, and at that moment a great cheer rose from the gallery on the bank. The fish weighed seventeen pounds and we have a group photograph with Ruth beside it where she and the fish look about the same size.

As the years followed we would fish together almost daily in the afternoons after my writing; it was a splendid period of our lives. We always seemed to enjoy one another's company, and the setting on the river could not have been more beautiful. I remember that on one perfect sunny afternoon in a secluded spot we were glowing and suddenly became overcome with desire. But as we progressed it soon became obvious that lovemaking in high waders presented insurmountable problems, so finally we just lay there in the grass laughing at the absurdity of it all.

For many years writing in the morning and fishing (or walking) in the afternoon has been, and continues to be, my way of living on the Margaree. The only thing that has changed is that I no longer send illegible handwritten manuscripts to a secretary in Princeton, all spliced with cutting and pasting. Now (and that includes these very words) everything pours into my laptop computer and emerges looking as though it had already been published.

I am often asked how I can write in such a remote spot without a reference library. I try to bring up what books and articles I might need but inevitably I fail. Sometimes I can send for something that is missing. More often I will rely on my faulty memory, and then check everything when I get back to Princeton. It is as though in the middle of a very factual work, one suddenly introduces a bit of wild fiction, for I am never quite certain if I have remembered correctly. This got me into trouble only once. I was writing a paper in which I wanted to refer to the work of a Japanese scientist, but I did not have the reference. So I gave him a fictional Japanese-sounding name: Okimoto. When I got back to the library I unearthed the correct name (which I can no longer remember—I can only remember my Okimoto) and changed it in my paper. But I did not root out all the Okimotos; I missed some so that when the paper was published I received a couple of queries about the work of Okimoto.

The other reason that Cape Breton has become such a splendid place to write is because everything is so beautiful. In the beginning of each summer I am tremendously keen to have those wonderfully uninterrupted mornings where I can think about one thing: what I am writing at the time. By the end of the summer I yearn to get back to laboratory work and the bustle of university life.

The only time I have varied my summer schedule was when Dr. Park was very old. He was a good friend whom I liked and admired greatly; he had long retired and spent the summers on the Margaree alone in his cottage. He could not drive, so every Thursday we would go fishing together. Those days are etched in my mind. He had so many delightful stories he would rarely repeat himself. One day he told me that some years ago he received a telephone call from New York from the president of a well-known company that made fishing equipment; the man wanted to come to Baltimore to consult him. He said, "You know, John, I have always tried to appear diffident about my science and medicine, but I cannot conceal my absurd vanity when it comes to my fishing, so I was thrilled at the prospect of his visit. He came to the pediatrics department and my secretary ushered him in. After the formalities and when he was all settled, I asked him what I could do for him. He replied. 'My feet hurt.'" The pathos of the moment was not lost on me, and again came that wonderful smile.

We continued these expeditions during the summer of his ninetieth year, but he was not well. He had a pacemaker by then, and we would worry about him all alone in the cottage. One day he had a bad spell and mutual friends and I decided I should spend the night with him and see him through this rough patch. When I told him what I was going to do he said absolutely not; in fact, he was quite irritated at the thought of this invasion of his privacy. The next morning the telephone rang while I was at breakfast, and that familiar voice said, "This is Ned Park telephoning. You will be happy to hear that I survived the night."

My last letter from him was in March of 1969, his ninety-first year. In it he says,

I was completely surprised to receive your new book. Perhaps I ought not to have been surprised, for, it seems to me, no sooner do you reach the Margaree than you become pregnant. Your fishing clothes hide your state and no one suspects, when suddenly a book is born.

The last twenty years of the century showed a steady increase in the progress of biology. There may not have been so many radically new discoveries and turnings into new directions as there had been earlier, but there has been a great deal of important solidification and refinement of the advances of the past. It is more difficult to look at this recent period with as much objectivity as the earlier ones, simply because it is so close. One thing is clear: the number of biologists has greatly increased, along with the number of new journals. Again the largest proliferation has been in molecular biology, but it is generally true for all of biology, including ecology because of the very real concerns with the future of our environment.

We have finally come to the point where the prevalence of people is beginning to be felt: the forests are disappearing, the suburbs are expanding, the traffic is increasingly heavy over our ever-expanding roads, the congestion in airports is ever more evident, and wherever we go we see the problems of pollution and water shortages. In some regions of the world famine is a serious matter as is the decimation caused by AIDS, especially in Africa, but disease and hunger have always cursed mankind; they just take different forms today and certainly do not curb the overall population growth. Along with this rising multitude of people have come amazing technical innovations: there may be too many people, but communicating with one another is now much easier what with e-mail and cell phones, and getting to distant (or near) places is quicker despite the jams. The world is simultaneously becoming larger and smaller. In many countries this has been a period of prosperity and there have been no widespread world wars.

*   *   *

In biology there have been some important advances on a number of fronts. The power of mathematical modeling in ecology and evolutionary studies has gone forward at a brisk pace with rewarding results. There has been great activity in neurobiology and the study of the brain, involving again mathematical modeling to gain insights into its complexity, as well as a greater knowledge of how the neurons communicate with one another on a molecular level. The progress in our understanding of the relation of genes to behavior has been dramatic. Developmental biology is making great strides forward for, thanks to the pioneer work of Christiane Nüsslein-Volhard and Eric Wieschaus and later many others, it is possible to find the genes responsible for developmental steps; now we know an enormous amount more of the chemical details of development in both plants and animals (and slime molds!)—note again it is those fusions with genetics that generate all sorts of important new things.

One striking aspect of the study of both genetics and development is the emergence of "model" organisms. This has occurred because the initial experiments were done on a particular plant or animal, and with that head start the organism becomes the logical focus for subsequent research. In this century there has been a tremendous emphasis on the bacterium *E. coli,* the fruit fly *(Drosophila),* and the nematode worm *(Caenorhabditis),* but beginning back earlier in the twentieth century and further back into the nineteenth century there are many others that have played a small role. To mention a few, there are Mendel's garden peas, followed by other organisms such as maize, amphibians, chick and sea urchin embryos, yeast, myxobacteria, zebra fish, the higher plant *Arabidopsis,* and cellular slime molds. One could add a few more and the list would still be incomplete: for instance, ciliate protozoa, *Hydra* and other hydroids, sponges, *Volvox* and other algae, true slime molds *(Myxomycetes),* fungi *(Phycomyces),* mice, and other mammals. The degree to which these various examples have been directly illuminated by genetics and molecular biology varies, but even in those cases where the influence has been small (due to the lack of attention) this is beginning to change. In fact, one can say that it is inconceivable to do developmental biology today on any organism without genetics and molecular biology.

Evolutionary biology has also evolved. Our appreciation of natural selection operating at all levels, from genes to groups of individuals, has deepened, with a greater understanding of the consequences. We have become more appreciative of how the environment interacts with the genes and how changes that are not controlled by the genes play a key role in evolution. To me one of the most interesting phenomena is that Charles Darwin is more esteemed today than he has ever been. This has happened not only because his views fit so well with our modern ones but also because many a time someone imagines they have a new insight into some detail, only to discover that Darwin had the same insight well over a hundred years ago.

One of the astounding things to me was the speed with which new equipment, new methods, and new tools developed into a great crescendo by the end of the century. To list them all would paralyze my tale—there are so many major advances—but let me mention a very few well-known high-lights, some of which affect our lives. The PCR method of isolating small bits of DNA and amplifying them to make many copies has become a vital tool in forensic medicine and has decided the guilt or innocence of a criminal. It is also used to great profit in molecular biology and even in evolutionary biology, for now we can trace the ancestry, the phylogeny, of any group of animals or plants. (It has even played a role for the modern historian, for instance, by showing that Thomas Jefferson may have been the father of his slave Sally Hemings' children.)

In the world of microscopy the old brass microscopes have been replaced by the most extraordinary computer-driven microscopes that have revealed the hitherto unseen interior of living cells and tissues in ways that go beyond the imagination.

Behind all these tools of biochemistry and cell biology, and behind so much that we do today, is the computer. Computers have introduced not just the conveniences of word processing and e-mail, wonders that they are, but they have become the way we control and operate our complex equipment in the laboratory. Furthermore, they have become the way to store vast quantities of data (and retrieve instantly one item from a great multitude). And I have not touched upon their role in mathematical modeling. We are in an era where everything gets a mathematical model, sometimes to great benefit, for it gives us insight where it was lacking. A pre-computer example of how effective mathematical insight can be is the work of Robert MacArthur and E. O. Wilson on island biogeography: the mathematics illuminated why and how the fauna and flora of islands differ from those of the mainland. In more current models the computer plays a key role, and often the objective is not to simplify but to predict how future changes would affect the complicated present.

The computer also deals with the biological literature in a wonderful way. When I was a graduate student it was expected of all of us that we would have a good idea of the important papers in every journal in the "current" rack in the library. Today there are so many journals (and they are so specialized) such familiarity would be out of the question. Fortunately all these journals are online, and with a click or two we can access an abstract and often the entire text (and the figures in color). Not only that, but we can search for a subject or an author with instantaneous results. The good part is that a vast literature is within our reach; the bad part is that we become ever more narrow and specialized.

In my own research, as I mentioned earlier, among other things I wanted to find out the nature of the repelling gas that the slime molds used, and this turned out to be ammonia. I sent the paper to the journal *Nature,* probably the most difficult journal in which to get a contribution accepted, and after some time they agreed to publish it. They sent me the proofs from London, but for some reason they came in two envelopes. It turned out the second envelope had been sent to me by mistake; it should have gone to the Washington, D.C., office of *Nature.* Being partly human I read the contents, which turned out to be all the interoffice memoranda concerning my paper—quite fascinating. Among other things they kept referring to me as the GOM of slime molds. I was no longer a Young Turk, but Grand Old Man seemed to me going too far.

It was an extraordinary bit of luck that the solution to the chemical nature of acrasin should have come so easily. As I described earlier, one of the big projects in our laboratory in the 1970s had been finding the chemoattractant or acrasin that gathered the amoebae together into a multicellular mass. Not all slime molds used cyclic AMP for this purpose, so some years later I started a major effort to find the chemical composition of the acrasin of one of the other major groups of slime molds. This took ten years, by contrast to the practically overnight cyclic AMP success story. The solution was largely due to the heroic efforts of my research assistant Hannah Suthers, who collected and concentrated gallons of liquid that had surrounded aggregating amoebae. There were a number of people who contributed to the early phases of this project, and our final bit of good luck came when Osamu Shimomura, a distinguished chemist then at Princeton, agreed to do the ultimate purification. With his unparalleled skill he went through a whole series of purification steps, so that ultimately he had a minute quantity of material that was 98 percent pure. He sent it off to an outfit that does mass spectrographic analyses, and from their result it was possible to identify a new compound that was rather a unique combination of two amino acids, which we called "glorin." The compound then had to be synthesized, which was done by some chemists who specialized in such syntheses, and they finally sent us a few grams of the presumed synthetic acrasin.

It arrived in the afternoon and Hannah Suthers said she immediately wanted to use the amoebae that were ready, which meant staying late in the laboratory. I had to go to a retirement party for a dean, but periodically during the speeches I would desert Ruth and go to a telephone booth to call Hannah. As the evening progressed the manufactured compound slowly revealed itself—it met our highest hopes: it was active at the predicted, very low concentrations. I was so excited that, during the course of the dinner, between my telephone calls, without realizing it I must have consumed a large amount of wine. I had no sensation of it at the time, but the next morning I had the most impressive hangover, worthy of a gigantic drunk. The entire excitement was almost as keen as the arrival of a new child, but in this case the pregnancy had lasted ten years.

During the 1980s I devoted much energy, in Scotland and elsewhere, to what I hoped would be a major work. My plan was partly to penetrate more deeply into the relation between those biological phenomena that were under the immediate direction of the genes, and those in which there were many intervening steps between the initial gene product and the end result, such as genes that affect behavior. At the same time I wanted to relate those ideas to another general theme that always fascinated me. The first organisms on earth were small and relatively simple, but during the course of evolution both animals and plants became larger and more complex, a fact amply recorded in the fossil record. My aim was to show how this could be accounted for by natural selection. Furthermore, ever since my exposure to the influence of Robert MacArthur I wanted to understand how these evolutionary changes in the structure of organisms might have influenced, or been influenced by, their ecological environment. I wove in all my previous ideas on the importance of organisms as life cycles and on the relation between a selection for an increase in the size of organisms and the concomitant adjustments that needed to be made to remain efficient, such as an increased division of labor within organisms to accommodate the size. I tried hard to put all these themes together in one place and show the connection between them in *The Evolution of Complexity*, which was published in 1988, and later in a lighter and more personal account in *Life Cycles*, published in 1995.

My sabbatical and summer writings over the years were cumulative. The ideas were a succession that seemed to want to refine themselves; it is as though it took me forty years to think a problem through. I have always flinched when I hear the old adage (I no longer remember who said it) that each author has only one book in him, and all subsequent books are the same book in different clothing. For myself, I like to think that I am just a slow thinker and a slow developer, and that the one book I have inside me has taken forty years of larval development. Think what might happen if I live to be a hundred!

As I look over the new ideas that I developed in my books, I see that some have mercifully died while others flourish today in the literature without any discernible connecting thread to those earlier themes of mine. This was no doubt due partly to their being ideas that were "in the air," but partly this was my own fault because I either used labels that did not properly convey the concepts, or I put the matter in a context that was too restricted. Let me give two examples.

One is the concept that we know today as modularity. It is the idea that in development the genes and their products are packaged in relatively independent units. As a result they can change, say, through the mutations of some genes that will only affect their module and not other parts of the body. In this way constructive or advantageous changes can accumulate with less chance that they will have deleterious effects on the rest of the developing embryo. This concept is implied in old ideas that began in the nineteenth century with Ernst Haeckel and were reexamined and honed by many others in the twentieth century. It is known as heterochrony, where different organs within the body can alter their time of appearance so that an organ that arises earlier in the development of an ancestor appears late in the development of a descendant. In another form it is also the subject of a famous essay by Herbert Simon, who pointed out that a watchmaker could increase his efficiency if he first put the parts together in smaller units instead of putting the watch together piece by piece. My early mistake was to identify the problem but give it the forgettable label "chains of steps" in *Size and Cycle.* Years later I came up with "gene nets" in *The Evolution of Complexity,* which was a slight improvement. By then the computer age was upon us and the concept was well established: information could be effectively stored in "modules." This has become the obvious and best way to label an important biological phenomenon.

The other example centers around variation. Classically variation is always assumed to be genetic, but from the beginning it has been appreciated that there can be variation induced by the environment. Indeed, this is the underpinning of the fruitless nature-nurture conundrum. My thought, which stemmed from work on slime molds, was that there was a third kind of variation that was neither genetic nor environmental, but a variation due to chance, and that it might be put to use in development. My slime mold amoebae varied in diverse ways, such as size or stored energy derived from when they stopped eating, and those at one end of the spectrum of these characteristics tended to become stalk cells while the others tended to become spores. I called this stochastic variation "range variation" in *Size and Cycle,* and I can remember being chided for the idea by a distinguished population geneticist. But now the idea has reemerged in an important book, *Chance, Development, and Aging,* by C. E. Finch and T. B. L. Kirkwood, who point out that aging shows the same kind of variation. Genetically identical animals do not die on the same day, but vary enormously in their life span, even in carefully controlled, identical environments. Finch and Kirkwood are the ones who had the wit to call this "chance variation," so much more compelling than "range variation."

I gave my last lecture to the freshmen and sophomores in the general biology course in the spring of 1990; I was just retiring and had reached the age of seventy. Fortunately someone had a tape recorder and gave me the tape of it because otherwise I would have forgotten what I said. It is a bit difficult to hear my words through all the rustling and coughing of the students and many of my old colleagues and friends who came to wish me well. I confessed that forty years earlier, when I gave my first lecture in the course, I was incredibly nervous, and after all those years I found myself just as nervous at my last lecture. I did have more problems in the beginning, however—I remember one student who consistently sat in the front row sound asleep with his head back and his mouth wide open. I resisted the tremendous temptation to drop a piece of chalk into it. Now all the sleeping people sit way in the back. That's progress.

The whole event warmed my heart. More than anything it made me feel that I was entering a new era, and I did not know quite what to expect. There were two very good consequences: I no longer had to correct any exams, and I decided never to write another proposal for a research grant, both of which I consider regressive activities. What I regretted was that I had always enjoyed my teaching and expected to miss it. On the other hand, it dawned on me rather quickly that being emeritus was going to be like a continuing sabbatical, and I had always found leaves of absence unrivaled periods for scholarly activity.

I consider myself very lucky to be able to keep my old office, which I have occupied since 1948, and in addition to be given some mini-lab space among the graduate students where I am able to continue with experiments. The office has a wonderful antique appearance, not much changed since it was built in 1910. It still has four lamps hanging from the ceiling with conical, green glass shades. It makes me feel as though I should be wearing elastic armbands and a celluloid eyeshade.

One great advantage in being emeritus was that I could travel at any time of the year, and did not have to ask the dean for permission. Going places was something both Ruth and I enjoyed. It had always been my firm belief that the best way to travel was to go to one place and stay there for a while rather than rush about. By a stroke of good fortune in 1990 I had an invitation from a friend in Bangalore to join a conference he was organizing at an old hill station called Pachmarhi in central India. The prospect was exciting, but unfortunately I had no funds for the trip. My friend said hold on a bit—he would see what he could do. Some time later I got a letter from the president of the Indian Academy of Sciences asking me to accept an appointment as the Raman Professor for a period of two to three months, and not only would that cover our travel but I would receive a stipend while in India. Furthermore, I could have space at the Indian Institute of Science and either live there in the guest house or at the Raman Institute, which was not too far away. It did not take me long to accept and we made our reservations to leave the coming October.

In years past Ruth and I had talked about visiting India, but in our ignorance it was low on our list because we feared the extreme poverty and felt that unless one was Mother Teresa, it was a place to avoid. How wrong we were, and how glad I am that circumstances overruled worries that were spawned by ignorance. For us it was a new world. In some ways it was like starting life all over again. I do not mean to imply that extreme poverty is not prevalent. We were warned that near the Bombay airport we would see and smell extensive shanty towns and we must steel ourselves beforehand. Being forewarned helped but only a bit.

Our arrival at the airport was rather grandly received by an airline official who whisked us through customs, helped us change some money, and got us into a taxi for the hotel, brushing off the swarm of young boys and men who wanted to carry our bags. It was our first inkling that a Raman

Professor was something quite special. Very early the next morning we flew to Bhopal on a local flight and right away we knew we really were in India. There was a simplicity and homeliness to everything in the domestic airport that stood in stark contrast to the international airline and the fancy hotel. In Bhopal all the conferees met in a small hotel for a delicious and much-needed breakfast and then boarded a battered bus for the most amazing all-day ride to the hills of Pachmarhi.

The seats in the bus were hard and uncomfortable but we hardly noticed, for the sights we saw on the roadside were so fascinating. There were the endless fields with workers leaning over and tending the rice or doing other chores. The women were all in colorful saris, which gave an impression of unreality. All along the roadside there were more men and women walking, often carrying loads. Occasionally we would see a painted oxcart with big wheels, all yellow and red and blue—even the horns of the ox might be gaudily painted as well. Often in the fields, or along the road edge, were cattle, but unfamiliar to us was the sight of buffalo with their curious black skin that looked as though it had been polished. Now and then around the base of a tree by the roadside there would be a small band of bonnet (macaque) monkeys, playing and watching the people and the traffic going by. The only jarring note in these riveting pastoral scenes that whizzed by us was that our driver spent most of his time leaning heavily on his horn for very little reason that we could see. It was as though he wanted the world to be aware of our presence, and in that he no doubt succeeded.

That was not the end of the marvels. The villages were a beehive of fascinating activity: dogs and gray pigs wandering about the streets, and minute shops, some seemingly no larger than a big box with a man sitting cross-legged inside it. Everyone seemed to have clean clothes; the men in very white shirts, the women in dazzling saris, and the children in school uniforms—one of the many unresented leftovers of British rule.

We stopped at Itarsi for lunch. It is a small railway town and as we staggered out of the bus into the intense sun, there immediately was the question of where twenty or so people could possibly find a place to eat. I need not have worried because my old friend and fellow slime mold biologist, Vidya Nanjundiah, who was the organizer of the conference, disappeared down the street and soon returned to say all was set. We followed him to a small, dark, and dingy restaurant where he had already ordered our lunch. It was quite delicious—spicy rice and a mixture of vegetables, along with a cool beer. We were ready for a hot afternoon of pounding travel.

Pachmarhi is a pretty town on the top of some hills. It used to be a base for the British army, and now it is where the Indian army has its camp for training army bands. We were put up in a small inn, and sometimes we could hear military music in the great distance. Our quarters were comfortable, the beds incredibly hard, which we liked, and the pillows no thicker than an empty envelope, which we liked less. But best of all was the opportunity to meet many Indian scientists, both senior people and numerous bright and engaging graduate students. There were almost as many women in the group as men. Everyone was wonderfully open and responsive so that we had many lively discussions. There were a few know-it-alls in the group, but they were not rebuffed or in any way treated as though they were different. Indians are perfectly equipped for lectures and conferences. They do not hesitate to give their views, and that is true of the students as well; they are responsive and interested in what others have to say, and they are invariably considerate and polite. Furthermore, they laughed at my jokes when I gave my lecture; what more can one ask of anyone!

Pachmarhi is in the midst of a forest, which allowed me to see some Indian wildlife. There was my first encounter with langurs, which I found very exciting. They have a distinctive hoot that carries great distances. For monkeys, they seem large as they occasionally lope along the road. They are not particularly disturbed by the presence of human beings; they give the impression of looking down on one as some sort of inferior primate, and they might well be right. More often my excitement centered around birds: I saw my first hornbills and my first bulbuls—even the ever-present mynahs were a source of fascination at first. Perhaps my biggest surprise was seeing wild peacocks and peahens scurrying through the edge of the woods.

The trip down to Bangalore was another big new adventure. We were driven back to Itarsi and waited for the train on a very hot platform, alongside a cow that was calmly eating a newspaper out of the trash bin. The only fright was that Ruth suddenly fainted, but fortunately I caught her on the way down. We stretched her out on our suitcases and she soon revived, recovering completely after a good sleep on the train. There were six of us: besides my host Vidya Nanjundiah, there was a mutual friend from Japan and two graduate students of Vidya's whom we got to know well. The compartment consisted of four berths perpendicular to the train, and two parallel and across the aisle. They were simple pallets covered with a sheet and one of those modest pillows. We spent most of our time peering out of the windows, with quite a different view of the countryside from that of the bus but one equally absorbing. The trip took thirty-six hours, and we were very lucky to be under the care of our Indian friends because they knew how to get the meals (which were remarkably good for train fare) and how to dash out at station stops and buy some of those delicious, small bananas without being left behind.

Our life in Bangalore could not have been more pleasant. We lived in the guest house and my office was a five-minute walk away. The grounds of the Indian Institute of Science are like a great botanic garden, with a large variety of flowering trees and shrubs, all beautifully cared for by a team of gardeners. There were a great number of colorful birds from large vultures that nested in the main tower of the Institute to small sunbirds, a pair of which nested in a ball of old caterpillar cobwebs by the entrance of the guest house. Dear to my heart were the coppersmith barbets, the minivets, and the flocks of parakeets that looked like small parrots. But there were many other sights as well, such as a sudden cloud of huge fruit bats, or flying foxes, that would appear in the very brief tropical twilight. I would never leave our room without a pair of binoculars.

Mainly through Vidya, but also through the Raman connection, we met many splendid people, very bright and interesting to talk to and always kind and helpful. We were taken to the Indian Academy of Science annual meeting, which that year was held in Bhubaneswar on the east coast of India, where I had to give a lecture. It meant more fascinating and different countryside, but it also meant meeting more interesting people. We had two guides, one of whom was Raghavendra Gadagkar, who took care of us from Bangalore and back. We became good friends and went on to collaborate on a paper that compared his social wasps and my social amoebae. At the meeting a graduate student I knew from Bangalore asked me if I was enjoying myself in India. I said enormously; I could not get over or explain why everyone was so kind to us. He said, with a big smile, "That's because in India we respect learning and we respect age."

I realize that in any country there is a mixture of different degrees of excellence in science and that I was at one of the centers of the very best science in India. Many of my colleagues there were world-class. Generally they had made their postgraduate studies in a foreign university, they all spoke perfect English—now the universal language of science—they were broadly literate outside their specialized field, and their manners, more often than not, were warm and charming. To give some examples, Gadagkar, whom I mentioned above, has done (and is doing) work of great importance on the evolution and behavior of social wasps; Sukumar, who has done so much to elucidate the habits and the ecology of wild elephants; Nanjundiah, who is one of the leaders in my own field of slime mold biology.

The other thing that is impressive is the high level of intelligence and achievement of the graduate students. They are so easy to talk to, so responsive, and the best are not afraid to use their imagination. The whole atmosphere of the Indian Institute of Science vibrated and I found it very stimulating.

One day at lunch in the guest house I sat next to an interesting physicist from another part of India. He was very forthright—no beating around the bush—and he asked me if I did not find the social hierarchy in India disturbing, coming as I did from a relatively democratic country? It dawned on me that I had never thought about it, which I confessed, and I immediately began to take notice. In the department where I was working there indeed was a remarkably clear-cut hierarchy. The head secretary was a nice young man from Goa, and I realized that when I went to him for a pad or some paper clips, he never gave them to me himself; he always gave a rather gruff command to a subordinate who scurried off to get what I needed. Beneath the subsecretary were the cleaners who swept out the offices, and they were ordered about by everyone. I mentioned to a charming professor at the Raman Institute that I noticed in the horrendous traffic of Bangalore that the only traffic rule was one of might. The trucks cut in front of the cars, the cars in front of the motorcycles, who lorded it over the bicycles. He laughed and said that is the Indian hierarchical way.

During one period I could watch some major construction work going on in a building across the way from my office. The chain of command was obvious. There were the big bosses who arrived in clean white shirts and would give orders with abandon; there were the foremen who oversaw the workmen; there were the masons next in line; and finally there were many men and even more women who did the manual labor. The women worked nonstop for a very long day, carrying great loads of bricks in a large saucer on their heads and walking barefoot up a steep bamboo ramp to the second floor, where they would dump the bricks in a pile in front of a man who threw the bricks one by one to another man on the third floor, where they were put on another big saucer, and carried by another woman (always in a sari) to the mason on another part of the roof. Cement was made in a hand-cranked mixer (two men cranking), and once the cement was in a saucer it was handed with great speed from one man above another on a long ladder to the roof. Needless to say, all this interfered greatly with my work. The thing I found especially distressing was that the women, who always seemed to me to be working hardest, were paid very much less than the men. Yet despite all the extremely hard work they did not behave as though they were oppressed: I could see them joking and laughing as they pushed on with their heavy labor.

Middle-class Indians lead a comfortable life sometimes bordering on the elegant. I can remember going to a dinner at the house of a professor who was a kind friend. We were the only non-Indians there. The house was simple yet quite beautiful. The living room had a dark red stone floor, white walls, and no pictures, just a small Hindu shrine high on one wall. There was no furniture at first, but as the food and drink arrived we were given chairs. It was a beautiful scene because all the women wore particularly elegant saris that made everything glitter.

I think I learned more about the people of India from reading the novels of R. K. Narayan than I did from my own limited view. His tales of the people in his small imaginary town give such a crystal-clear and sympathetic insight into the minds of his characters that one learns about humanity in general, and Indians in particular. Those short novels are gems. I started reading them there, and they suddenly made me see things that I had not the understanding or the wit to see before. We visited Mysore where he lives, and I wished we had met him.

One is constantly reminded of the enormous influence of the British. It is not just that English is the universal language of the educated, for there are so many local languages that there had to be some way all Indians could communicate with one another, something that Hindi was never able to achieve. There is much more. What surprised me so much is that even though the desire for freedom from British rule was and remains paramount, things British are not resented; in fact, the good points that they left behind are admired. Nothing brought this home more to me than the huge square in Bangalore where there is a gold statue of Mahatma Ghandi at one end walking with his beggar's stick, and at the other there is an enormously stout and severe Queen Victoria sitting unsmilingly on her throne. How remarkable it is that she was not pulled down in the joyous moment of independence. But then that was the moment Mountbatten, who was the viceroy at the ceremony where power was transferred, was mobbed as he and Lady Mountbatten rode in a parade in their carriage; the mob did not come to do harm, as was feared, but freed the horses and pulled the carriage of state themselves. The Indians are a proud people, but they do not bear the same kind of grudges I am used to.

For me the best moments of all in India were spent in wilder places. We saw many ancient Hindu temples that were spectacular with their filigree depictions of the great legends (or of the joys of sex), and while I found them beautiful and impressive, it was the wilds that held me in thrall. This was especially the case in our two visits to the great preserve of Mudumalai, south of Bangalore. The trip there is a lovely drive through the country; much of the road is lined with ancient trees put there by some maharaja many years ago. Mudumalai is a great protected area with a fairly dry scrub forest of teak and many other trees. The wildlife is wonderfully abundant. The insects, the birds, and the mammals all competed for our attention. We were particularly fortunate in having as our adviser and guide the distinguished elephant authority Sukumar, a friend from the Indian Institute of Science in Bangalore.

He took us one early morning on a bumpy ride in a jeep over small trails barred to all but the wardens. What a glorious morning—a jungle misty and fresh before the heat of the day. We saw so many things: many chital (which are small deer), numerous langurs, a great variety of birds, including an incredibly bright-colored kingfisher, and a fine sighting of a large sloth bear; but undoubtedly the most exciting were the encounters with the wild elephants. They are considered dangerous so we were not allowed out of the jeep, although Sukumar would cautiously sneak out to photograph them as part of his study. Later he gave us a large photograph of a group charging him while he was taking their picture—he said he made it back to his jeep just in time. Later we took a ride though the jungle on a tame elephant, which provided the great advantage of not only a high seat but no noise—it was like traveling in a balloon. Also, the other animals seemed to ignore us, and no doubt as a result we saw a sambar, a huge deer, and a small heard of gaur, shy wild oxen with white stockings and brown coats. Our elephant gave us a smooth, swaying ride. My only biological observation of note was that whenever the animal climbed up an incline, it farted with every step; I have not yet published this remarkable observation.

I could take short walks with my binoculars around the place we were staying, and those walks remain especially clear in my memory. On one of them I suddenly came out into an open place in the forest, only to realize it was open because it was a broad elephant trail with footprints and piles of elephant dung strewn here and there. I had the feeling I had accidentally entered another world, which is exactly what I had done.

Continuing with experiments has been a tremendous joy. Previously I had a laboratory assistant (occasionally two), usually at least one postdoctoral fellow or visitor, and some graduate students and seniors; now I was down to myself. It was like being a graduate student all over again, a period of my life I remember with fondness because I did my own experiments, with my own hands—the successes and the failures could be all mine. It was as though I had again been given the freedom of the cellar and could play with my pre-teen chemistry set.

There were some striking differences between those periods: in graduate student days all our culture dishes for growing slime molds were made of glass, and now they are throw-away plastic; no more massive dishwashing, and the plastic is better optically so one can see one's molds much better. The balances to weigh out chemicals are now far superior and easier to use, but I was embarrassed to confess that I did not know how to use them—I always had someone else to do it for me. I would furtively ask about how to do these elementary things, and soon I could again prepare my own culture media and the solutions needed for my experiments. Another big difference was my time-lapse equipment. As an undergraduate I had taken that film of slime molds I mentioned earlier with the most amazing, ancient time-lapse apparatus. It consisted of a gigantic collection of brass gears attached to a box of a movie camera that looked exactly like the hand-cranked cameras they used to film the Keystone Cops. The entire machine was so heavy, and vibrated so much, that it had to be on a separate table from the microscope and the optical connection between the two could not touch. Once I had taken a reel, I had to send it off to get developed and it was a week or so before I knew the result. The remarkable thing is that those early films are so good, but I had to work hard for the results. Now, in my mini-lab I have a video camera, a VCR with time-lapse, and a TV screen. I can shoot something all morning, and before I go to lunch I can play it back in a few minutes. Furthermore, there is no danger that the exposure is wrong; I record exactly what I see on the screen. Laboratory life is getting soft.

With this new toy in recent years I have been able to forge ahead onto some new ground. Some of the experiments on slime mold slug orientation that I mentioned earlier were done this new way. To give an example, I was interested in whether differences in ammonia produced in a migrating slug were sufficient to turn the slug. The ammonia in the slug presumably was produced by the breakdown of proteins, which is a common cellular process encouraged by digestive enzymes specific for that purpose. To prevent ammonia production, different inhibitors of these enzymes were tested by soaking them up in a small bead of gel, and then placing the bead on one side of the tip of the slug. The experiment gave a striking result: the slug turned toward, and wrapped itself around, the bead. In other words, the inhibitor stopped protein breakdown and ammonia production on one side of the slug, and as a result the slug curved around the bead. This was all recorded with the time-lapse video and could be shown speeded up dramatically on the screen. I was enjoying myself hugely doing my kind of experiments.

More recently I had a wonderful bit of good luck. I was trying to poke a fine glass rod through a migrating slug to see how the cells moved around it, but I did not have any glass rods so instead I pulled out some glass tubing to make a very fine needle of glass that had a lumen up its center (like a piece of miniature macaroni). I noticed that when I touched this small capillary to a slug, some cells moved up inside it by capillarity, the same way water will rise up into a small tube. I was curious to see what would happen to these cells, so I put the capillary under the microscope attached to the video camera. What I saw seemed very surprising: quickly two zones formed; the anterior one near the air bubble was smaller and its cells seemed tremendously active and moved about in great swirls, while the cells of the larger posterior zone appeared rounded and did not seem to move. This appeared to be a miniature replica of what one found in a larger, normal slug: the small anterior zone of active cells that will eventually make up the stalk (prestalk cells, as I had dubbed them many years ago) and the bigger posterior zone that will make up the spores (prespore cells). This capillary preparation has some great advantages. First of all, one could see the cells, which is virtually impossible in the larger, normal slugs; and second, the zones formed very quickly as though the conditions inside the tube hastened the differentiation of the cells.

I wrote this up and sent it to some friends whose opinion I value, and the response was unanimous. What I had found was very interesting, but they all had grave doubts that the cells I described were really prestalk and prespore cells; how could I be sure I was not describing death in a tube? Their polite disbelief was nonetheless encouraging, and they urged me to find some more definitive way of identifying the cells. One of these advisers was Jeff Williams in Britain who had done so much to develop a method of identifying cells. He found genes that were expressed in either the spore or the stalk pathway, and a reporter gene was attached to those genes so that when the specific prestalk or prespore protein is made, it simultaneously makes a marker protein that is visible. In this way it is possible to see the first appearance of a protein that is characteristic of only prestalk or only prespore cells. The first such markers made in this way by Jeff and his group could be seen by fixing and staining the reporter protein, but now it is possible to use a reporter gene the comes from jellyfish, and it produces a protein that is fluorescent and can be seen with a suitable microscope in living cells over a long period of time.

The advice to use such markers was excellent, but it involved molecular biology skills that were quite beyond me and my mini-lab. In discussing the matter with my friend and longtime colleague Ted Cox, he said that a marker for cells that differentiated into stalk had just been developed, and it was the one needed, for it was attached to the fluorescent protein; furthermore, he would be delighted to join forces with me to look into the problem. So I soon became involved with three people in his lab in what turned out to be a most fruitful collaboration. Our plan was to use his strain of amoebae with the fluorescent reporter protein to show if the stalk protein was turned on in the capillary, and watch where the fluorescence appeared in the two zones in the tube. If my conjecture that the anterior zone was a miniature prestalk zone was correct, then that is where the fluorescence should appear. We planned to observe this in the confocal microscope that reveals fluorescence and projects the image on a computer screen. It is a wonderfully high-tech machine that looks as though it came from the flight deck in *Star Trek*.

We all assembled in the small darkroom and put our capillary with the genetically engineered cells under the microscope. At first they looked slightly more fluorescent in the anterior zone, so our hopes leapt, and as the minutes ticked on, it was clear the anterior fluorescence was increasing. We ran the experiment for two hours and we kept returning to have a look and with each visit we became more and more excited: the stalk proteins were being made almost exclusively in the anterior zone. It was not one instant of euphoria but a whole morning of progressive euphoria. We all felt as though we were on a high with some long-lasting drug. We had to do many more things to test our result to be sure there was no mistake, but that morning we knew in our hearts that the cells in the capillaries were a microcosm of normal development, and might prove to be an important tool to study the control of differentiation in slime molds. For me every aspect of this project was rewarding—from the initial chance discovery to the collaboration with all those bright molecular people. I indeed felt as though I was a graduate student all over again.

More recently I found, again quite by accident, that occasionally a small group of cells will crawl out of the end of one of these capillaries and the cells will crawl between the mineral oil and the glass, often being one cell thick. The remarkable thing is that these appear to be minute slugs and one can see every cell. Normal slugs on the surface of a culture dish will be made up of around 500,000 amoebae, big ones being over a million cells, while these mini-slugs range from 100 to 2,000 amoebae. They crawl forward, having a clear anterior-posterior polarity, and using the markers described above we could show that they are truly divided into normal prestalk and prespore zones. It is possible to see all the cells as the slug crawls, and as a result we have for the first time a clear picture of the slug movement itself. The beginning of differentiation can take place in two dimensions, which means that we now have a direct window into normal development.

A sadness has hit me in recent years. Ruth has been struck by Alzheimer's disease. It started to be noticeable some time after our fiftieth wedding anniversary; her doctor noticed it before I did. At first I felt bad I had been so insensitive and unobservant, but now I realize that the first symptoms had been so slow and gradual in coming during the previous years that there never seemed to be any change. Things altered, and the changes became dramatic and more rapid. For some time she could no longer read though she had been a voracious reader. She no longer could remember anybody's name except family. She often asked, "Where are the children?," her mind having gone back to when the house was teeming. She began to have great difficulty telling a story—the words just do not come. Yet with all her confusion she had moments of remarkable insight, as she always had. She enjoyed company and followed the conversation with attention, occasionally interjecting a characteristically penetrating comment. I found myself unable to complain—we have had so many happy years together, and for a long time she clearly not only recognized me but could show loving warmth in her look. One night I was already in bed and she was climbing in. Suddenly she looked at me in a tired way and said, "What's your name?" I replied "John," and she said, "Oh good," and climbed happily into bed.

As one looks at the entire century it is striking that there are ever-changing interests in biology. Obviously this is to a large extent governed by the new discoveries that have occurred with such éclat and frequency over the years. Nevertheless it is striking to me the degree to which fads and fashions play an important role. They are indeed directly related to the new findings, for those findings set the stage for the next fashion. I doubt if any of the great new advances, new directions, were driven by the fashion of the time but rather the reverse: the new direction started the fashion. The genes on the chromosome started cytogenetics; population genetics caused a flurry of important work on speciation or how new species arise; the structure of the gene gave birth to molecular biology; the idea of genetic relatedness as one of the explanations of cooperation did much to launch sociobiology; and there are others, some of which have been mentioned on these pages. Once the discovery, the new direction, was recognized, there followed many studies to explore all the consequences; the fashion was set, and everyone, including the funding agencies, continued on course. This is very much in line with Thomas Kuhn's ideas in *The Structure of Scientific Revolutions:* he labeled the innovative step entering a new "paradigm" and described the period of fashion as "normal science." (Robert Mac-Arthur once told me that if I wanted to start a new revolution in biology, I should use Kuhn's famous book as a field guide. Robert did start a new paradigm but without the help of Kuhn.)

A current new fashion is evolution and development. It is certainly not a new subject and has a venerable history. Darwin himself understood the importance of one to the other and devoted considerable space to it. In the nineteenth century the flamboyant Ernst Haeckel also propounded his "biogenetic law," which says that ontogeny (development) recapitulates phylogeny (evolution); the stages of development often parallel the evolutionary succession so, for instance, we see an early stage in our own embryos in which we have gill slits and a tail, reflecting our descent from a fishlike ancestor. This "law" is a great oversimplification that throughout the twentieth century stimulated much fine work, and a number of people were able to show that the relationship was much more interesting and complicated. While this work continued at a modest pace, it never reached the status of a fashion.

There are others—for instance, Waddington—who were greatly concerned with evolution and development. They appreciated that what evolved was not just the adult but the developing embryo as well, and genetic changes by mutation could affect any stage of development. This approach did not really take off until Nüsslein-Volhard and Wieschaus found a way of studying the genes in a fruit fly that were active at different stages of embryonic development. This fruitful merger of molecular genetics and development produced (and continues to produce) an immense amount of important discoveries uncovering the molecular steps that are development.

Also, the merger provided a key on how to study the evolution of development in a molecular way, something that hitherto had been inconceivable. For example, a set of genes called HOX genes are strung along on a chromosome and they are known to be responsible for laying down the main anterior-to-posterior regions of the main axis of an animal (in the same sequence they are found on the chromosome). These genes were first discovered in fruit flies, but now this HOX family of genes is known to be present in some form in all bilateral animals from the most primitive invertebrates to mammals. This has provided an unparalleled opportunity to study the evolution of the genetic control of development. It is in part these recent successes that have made the evolution of development an "in" subject right now, and there is a large amount of exciting new work in progress in many laboratories. There is even a new journal appropriately called *Evolution and Development.*

Evolutionary biologists are simultaneously acquiring a fresh interest in development. They are taking advantage of all these molecular advances, and building on the groundwork laid down by Waddington and others, they are now asking deep questions about how the genes and the environment combine to produce evolution. More and more there is an appreciation that the plasticity of the shape and structure of the organism plays a much-neglected key role in evolutionary change, something that had been pushed to one side in our love affair with the gene.

In recent years there have been some pioneers who have pointed out that knowing the role of the genes in an organism goes only a small way in understanding the making of the organism. To say that if one has the genetic map one can build the organism may be the case for a virus, but it certainly is not so for an organism made up of cells. This is an old, commonsensical idea, but one that was dwelling under the rug because it seemed like a complex bother compared to the simplicity of what the genes are doing. That it was recognized by the beginning of the twentieth century and before is clear: the egg carried a huge bag of cytoplasm along with its genes. I can remember Conklin in his swan-song lecture to the embryology class at Woods Hole wagging his finger at the students and saying, "You have inherited more from your mother than your father." All that cytoplasm in the fertilized egg not only sends out messages of its own but influences what messages the new combination of parental genes sends out. And as development proceeds, the immediate gene products—the proteins they produce—are not the end stage. Those proteins may be enzymes that promote another reaction, and as we know from the biochemistry of cells there may be a cascade of reactions producing a vast number of microenvironments within the cell, each one of which may play a key role in the steps of development. And each step may have an effect on other steps. The gene products in the cytoplasm start a wave of actions and interactions of an amazingly complex nature that lead to the building of the adult. The awareness that one needs more than the genes to make a fruit fly or a human being has again come to the fore, albeit in a more sophisticated form than we had a hundred years ago.

I ask myself where do I fit in all this? Ever since those lectures I gave in London in the 1950s I have been pointing out the essence of the previous paragraph. One could have the important events of development many removes from the initial gene products. In fact, in those lectures I argued that behavior in animals was one of the most distal (and most interesting) results of gene actions, and that because behavior is so far removed from the genes, it can become to varying degrees independent—it can take on a life of its own.

I know now that my perspective on evolution and development comes from my study of the slime molds. I said earlier that I worked with ciliates in France so that I would not become narrow and warped by just studying one peculiar organism, but I need not have worried (although ciliates are fascinating). Looking at the world through the eyes of a slime mold has not been limiting—just the opposite. By their very peculiarity they made me see other "normal" organisms in a new light. It was through them that I began to realize that organisms were not just adults but life cycles. That led me further to the idea that each life cycle had a point when it is a small, single cell, a spore or a fertilized egg, and at the other end of the cycle it became a large adult—and that trajectory is development. The duration and magnitude of the cycle correlate directly with the ultimate size of the adult. Adults come small and large, and all the sizes in between, and this raised many interesting questions, both physiological and structural, on the relation between increase in size and the division of labor. With division of labor comes complexity; in *The Evolution of Complexity* I argue that there has been a simultaneous selection for size increase at the upper end of the size scale, and there is always selection pressure for physiological efficiency, with the result that larger animals and plants have a greater division of labor and therefore are more complex.

In these aspects of the problem I was undoubtedly influenced by D'Arcy Thompson. He saw the importance of physical forces as constraining the construction of animals and plants of different sizes, and he scoffed at the idea that genes had anything to do with it. Today we can encompass both the genetic and the nongenetic elements in the construction of living things. His stress on the physical forces that cannot be ignored or circumvented is quite in keeping with our view today that genes, vital though they be, are not everything.

In our study of development and its evolution we see a great increase in complexity. This started right from the beginning, for a single cell is fantastically complex with all its myriad of proteins and all those other substances required for its functioning, and it contains an enormous number of different structures that are essential for its operation. Every aspect of the living is surrounded by or made up of huge complexity: think of the brain with all its neurons or any ecosystem with its fantastic diversity of plants and animals.

Clearly one wants to simplify this complexity, and in fact there are a large number of people who are approaching these problems using mathematics. Essentially they are trying to do exactly what MacArthur did for ecology: trying to analyze a complex problem by the use of simplifying mathematical models. There are even some institutions dedicated to the problem, with many individuals who are attempting to master all complexities, biological and otherwise, to find a way to sort out the strands and find the hidden simple understructure.

My aim has always been the same but my approach has been totally idiosyncratic. That may partly be explained by my severe limitations in mathematics, although I think there is another, more fundamental, reason: I want to do things my way. The very reason for my picking slime molds as an organism to study development was that they are primitive and relatively simple—only two cell types, stalk cells and spores, compared to the hundreds in our own bodies. Every organ inside us, heart, brain, liver, kidney, and so forth, is each made up of a great multitude of cell types. The hope was that by studying slime molds one might more easily find the basic elements of development. To some extent this hope has been fulfilled but certainly not to the degree I had initially imagined. In the first place they are made up of cells, and as I just said, each cell is a wonder of complexity. And also one finds in any biological problem, especially with our new molecular insights, that what may have seemed initially simple is not so: the deeper one digs for detailed facts, the greater the maze of complexities emerge. Before too long we will know the entire genome of slime molds, and it is already obvious that each step in their development involves a great multitude of mini-steps. How can one circumvent this problem?

It occurred to me that one could do it in theory, that is, by what physicists call a "thought experiment." The difficulty is that the aggregation of amoebae to form a slime mold must have first originated many millions, perhaps a billion years ago. During all that time natural selection has been operating with the result that every step has been reinforced with extra steps to make the development more reliable, even under variable conditions. This and the fact that they consist of complex cells undoubtedly are the reasons that present-day slime molds are not as simple as I had initially hoped.

To dodge these problems, one might ask what might have been the first extracellular signals used by the original ancestral slime mold. This in turn poses the question: what minimum between-cell signals are needed to produce aggregation and the cell differentiation into stalk and spore cells? The solution to this thought experiment would give us some idea of how slime molds might have started their social existence way back. But since the tracks of those initial events have been covered by masses of fine-tuning additions over those millions of years of natural selection, how can we possibly know what they might have been like? The idea I develop in my most recent book *(First Signals)* is that the only way one can get some insight into this question is by using a mathematical model that asks what are the minimal between-cell signals needed to produce slime mold development? Only such an approach will give us an idea of the fundamental nature of development shed of all the added frills.

Note that this book is another step in my life cycle. The seeds that produced it can be found in my early interest in the life cycles of lower plants, in my love for the particular one of slime molds, in my realization how size increase begets complexity and lengthens the cycle, in my admira-

COMING TOGETHER

tion of how MacArthur showed the power of simplifying mathematical models, and how all these seeds in my brain spawned a desire to understand how the first and fundamental multicellular life cycles arose. Perhaps this is another example of how I have taken a lone course that goes against the current general trend in evolutionary developmental biology. By sticking to the biological problems in both my research and writing I have managed consistently to keep out of fashion, and as the reader must have perceived, I rather enjoy the position of being a biological maverick.

My physical life cycle from egg onwards has been no different from that of any other animal. I grew, I differentiated into many cell types, and then I reached a lengthy physical plateau. (There was a shock some ten years ago when at my annual physical examination the nurse announced that I had lost an inch and a half in height since the previous year. I rapidly could see that in no time I would be a midget. Fortunately the next year a new nurse gave me back that one-and-a-half inches.) Human beings, once mature, stay pretty much the same height, but some other animals, such as fish, turtles, crocodiles, and even some mammals such as the elephant, continue to grow, albeit slowly, until they expire. My intellectual life cycle, which is the subject of much of these pages, has been very different from my corporal development. If at the egg stage I had any intellectual life I do not remember it! After it became recognizable in my preteen years it fortunately has not leveled off like the length of my bones, but has slowly and steadily risen over the years, just like the size of a turtle. I am told wisdom follows the same trajectory, but sometimes I have my doubts. Here, however, I am not talking about wisdom but the desire to think things through. A life is a life cycle.

Most of these words were written at my desk in our bedroom in Cape Breton. We have a very small wooden house in a very small village that we bought in 1964—the most fortunate decision we ever made. Over the years we have removed the linoleum floors and the layers of wallpaper so that one can now see the old boards. The first change we made was to convert the shed in back to a living room and put in a large window that looks out over the estuary of the river. We can see the mergansers, the cormorants, the herons, the gulls, the terns, and occasionally a bald eagle. Beyond the river is the small village of Belle Côte, and behind it rise the hills—it is a breathtaking view of which I never tire. The adjacent kitchen has an old wood stove that plays an important role on cold mornings. The old parlor in the front of the house now holds all the boots, the raincoats, the fishing rods, the clothes-washing and drying machines, and the piano.

From my desk the view looking up river is wonderfully distracting when the writing is having difficult moments— always changing with the time of day and the sun and the clouds. On a cool day I can smell the wood smoke from the kitchen. It always gives me the feeling that this is the place where I belong. It is as though I had the power to decide where to send my soul.

of slime molds, 57, 59–60, 87, 96; of ciliates, 104
Life Cycles, 181
Lindbergh, Charles, 30
Loewe, Otto, 14–15
Lorenz, Konrad, 85–86
Lowinsky, Clare, 117, 122
Lowinsky, Justin, 117, 122
Lowinsky, Ruth, 122
Lunt, Alfred, 27
Luria, Salvador, 82
Luxembourg Gardens, 110
Lwoff, André, 111–112

MacArthur, Robert, 155–160, 177, 180, 202, 206
Maestres, Ricardo, 141
Magic Flute, The, 109
Mallorca, 36, 170
Mangold, Hilda, 12, 79
Margaree valley, Cape Breton, 169–170, 172, 210
Marine Biological Laboratory, 84
Marx, Harpo, 27, 29–30
Marxist views, 154, 155
mathematical modeling, 177
McCollum, E. V., 138
Mendel, Gregor, 6
Mitchison, Dick, 129–130
Mitchison, Murdoch, 128
Mitchison, Naomi, 129–130
Mitchison, Rosalind, 128
"model" organisms, 175–176
modularity, 182
Moe, Henry Allen, 123
molecular biology, 83

molecular biology, at Princeton, 143–146
Monod, Jacques, 111–112
Morgan, Thomas Hunt, 7, 10–12, 38
Morphogenesis, 98–99
Mouse-trap, The (Christie), 117
Mudumalai, 194

Nanjundiah, Vidya, 188–190
Narayan, R. K., 192
National Science Foundation (NSF), 82, 91
Natural History Museum (London), 34
Nature (journal), 178
Needham, Joseph, 164
New Yorker magazine, 27
nightingales, 35
Nixon, Richard, 150
Nüsslein-Volhard, C, 175, 203

Officer Candidate School (OCS), 71–75
On Growth and Form, 17–20
Oppenheim, Paul, 96

Pachmarhi, 186–188
Panama, 61–64
Pardee, Arthur, 143
Park, Edwards A., 137–139, 172–173
Parker, Dorothy, 27
Parker, George, 9–10
Parpart, Arthur, 140
Perry, Lewis, 40
Phillips Exeter Academy, 38–41
Princeton, beginning at, 89–92

Racine (cf. to Shakespeare), 32
Rake's Progress, The, 109
Raman Institute, 185
Raper, Kenneth, 57, 58
Renn, Charles Easterday, 52
Roberts, Rebecca (daughter), 152
Rockefeller Foundation Fellowship, 102, 106
Rogers, Ginger, 27
Rosenbergs, Julius and Ethel, 114
Rothschild, Lord, 122
Roux, Wilhelm, 11

Sainte Chapelle, 108
Sainte Cucufa (park), 31
Sarton, George, 51
Science of Life, The, 34, 35
Scotland, 127–132
Second World War, 60, 64
Selfish Gene, The, 153
Shaffer, Brian, 161, 162, 164
Shimomura, Osamu, 179
Shroedinger, Erwin, 93
Size and Cycle, 161–162, 182
Smith, Kemp, 127
Spemann, Hans, 12, 79
Spurway, Helen, 118–119
Stadler, David, 92
Stadler, L. J., 93
Stehli, Emil (grandfather), 5–6, 25
Stravinsky, Igor, 109
Sturtevant, Alfred, 56
Sukumar, R., 194
Suthers, Hannah, 179
Swann, Michael, 131

Taylor, Wm. R., 46–48
teaching, at Princeton, 147–148
Tennessee Valley Authority, 52–54
Thatcher, Margaret, 132
Thomas, Charles, 68
Thomas, Thurlo, 41
Thompson, D'Arcy, 17–21, 206
Tinbergen, Niko, 85
Trotsky, Leon, 1

Ulva (sea lettuce), 47–49
University College (London), 116–119

Vaucresson, 31
Versailles, 108
Vietnam War, 133, 148–149; department reaction, at Princeton, 149–150
virus genetics, 82–83
Volvox, 125–126
von Baer, K. E., 41
von Frisch, Karl, 85

Waddington, C. H., 123–124, 202, 203
Wald, George, 55
Warbass, Dr., 8
Warner, Langdon, 29
Watson, James, 83, 144
Waugh, Evelyn, 34
Wells, G. P., 36, 116, 118
Wells, H. G., 28, 36, 37
West, Rebecca, 33
West-Eberhard, Mary Jane, 155